Y0-EFP-020

2-25-97

Philip J. Davis
Reuben Hersh
Elena Anne Marchisotto

The Companion Guide to
The Mathematical Experience
Study Edition

1995

Birkhäuser
Boston • Basel • Berlin

Philip J. Davis
Division of Applied Mathematics
Brown University
Providence, RI 02912

Elena Anne Marchisotto
Department of Mathematics
California State University, Northridge
Northridge, CA 91330-8313

Reuben Hersh
Department of Mathematics
and Statistics
University of New Mexico
Albuquerque, NM 87131

Printed on acid-free paper
© 1995 Birkhäuser

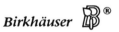

Copyright is not claimed for works of U.S. Government employees.
All rights reserved. No part of this publication may be reproduced, stored in a retrieval system, or transmitted, in any form or by any means, electronic, mechanical, photocopying, recording, or otherwise, without prior permission of the copyright owner.
Permission to photocopy for internal or personal use of specific clients is granted by Birkhäuser Boston for libraries and other users registered with the Copyright Clearance Center (CCC), provided that the base fee of $6.00 per copy, plus $0.20 per page is paid directly to CCC, 222 Rosewood Drive, Danvers, MA 01923, U.S.A. Special requests should be addressed directly to Birkhäuser Boston, 675 Massachusetts Avenue, Cambridge, MA 02139, U.S.A.
ISBN 0-8176-3849-0
ISBN 3-7643-3849-0

Text for The Companion Guide to the Mathematical Experience Study Edition was typeset by Martin Stock, Cambridge, MA
Printed and bound by Quinn-Woodbine, Woodbine, NJ
Printed in the U.S.A.

9 8 7 6 5 4 3 2 1

Contents

I. Introduction to this Companion Guide

   A Note to Instructors   3
   Special Features
     In the Text   5
       Topics to Explore, Essay Assignments,
       Problems, Computer Problems,
       Suggested Readings
     In the Companion Guide   6
       Expository Research Papers,
       Topics for Classroom Discussion,
       Projects, Tutorials

II. Chapter Guidelines

   1. The Mathematical Landscape   11
   2. Varieties of Mathematical Experience   12
   3. Outer Issues   32
   4. Inner Issues   37
   5. Selected Topics in Mathematics   54
   6. Teaching and Learning   59
   7. From Certainty to Fallibility   64
   8. Mathematical Reality   67

III. Sample Syllabus

   First Day Handout   75
   Syllabus   76

Contents

IV. Sample Group Activities

The Mathematical Experience	89
The Mathematical Landscape	90
Varieties of Mathematical Experience	91
Outer Issues: Utility	92
Inner Issues	93
The Pythagorean Theorem	94
Pythagorean Triples	95
Pascal's Triangle	96
Connections: The Golden Ratio and Fibonacci Numbers	97
From Certainty to Fallibility	98
Mathematical Reality	99

V. Sample Examinations

Sample Take-Home Examination	103
Sample Examination	105
Sample Midterm Examination	107
Sample Final Examination	108

VI. Topics for Expository Research Papers

Topics for Expository Research Papers	113
Suggestions for Grading Essays and Research Papers	119

The Companion Guide to
The Mathematical Experience
Study Edition

Part I
Introduction to this Companion Guide

A Note to Instructors

The first *Mathematical Experience* appeared in 1981. At that time, only a few years ago, it was commonly believed that it was impossible to make contemporary mathematics meaningful to the intelligent non-mathematician. Since then, dozens of popular books on contemporary mathematics have been published. James Gleick's *Chaos* was a long-run best seller. John Casti is producing a continuing series of such books.

In technology and invention, it's a commonplace that knowing what's possible is the most important ingredient of successful innovation. Perhaps the first *Mathematical Experience* changed people's idea about what's possible in exposition of advanced contemporary mathematics.

Alert readers recognized the book as a work of philosophy—a humanist philosophy of mathematics. It was far out, "maverick" (Philip Kitcher's term), virtually out of contact with official academic philosophy of mathematics. In the past 15 years, humanist philosophy of mathematics has bloomed. There are anthologies, symposia, a journal. The far-out maverick of 15 years ago might be the mainstream in a few years.

The first *Mathematical Experience* was a trade book, not a textbook. It was sold in book stores, not in professor's offices. But we heard over and over of college teachers using it, in the United States, Europe, Australia, Hong Kong, Israel. It's used in two different ways: "Math for liberal arts students" in colleges of art and science, and courses for future teachers, especially secondary math teachers, in colleges of education.

In mathematics teaching, it's a commonplace that "Mathematics isn't a spectator sport." You learn by doing, especially doing problems. Like all truisms, this is half true. Mathematics education as doing, doing, doing—no thinking, no conversation, no contemplation—can seem dreary. An artist isn't prohibited from occasional art appreciation—quite the contrary. You can't learn practical skill as a spectator, but you can learn good taste, among other things.

The first edition invited the reader to appreciate mathematics, contemplate it, participate in a conversation about it. It contained no problems. If a teacher selected it, he/she had to supply what the book lacked. The study edition will be more convenient for both teacher and student. It aims for balance between doing and thinking. There are plenty of problems, generous discussion guides, essay topics, and bibliographies. We've also introduced "projects": connected sequences of problems, rising in difficulty from very easy to a little less easy. They provide extra problem-solving enjoyment, and they make points about the nature of mathematics. We've written a section on differential and integral calculus—a complete course in 15 pages—and a section on the fascinating topic of complex numbers—fascinating from both mathematical and philosophical viewpoints.

The *Standards* of the National Council of Teachers of Mathematics appeared after the first *Mathematical Experience*. To a large extent, they validated our enterprise. We were following the *Standards* before they were written. The study edition does so even more than the first.

No longer are "critical thinking" and "problem solving" only features of mathematics. They've become catchwords in American classrooms. The second *Mathematical Experience* is a part of the dominant trend in American education.

Special Features

In the Text

Topics to Explore are listed in each chapter of the text. In this *Companion Guide* you will find you will find Suggested Topics for Classroom Discussions, Projects, and Tutorials that relate to the **Topics to Explore**. **Suggested Readings** in the text are resources for these topics, and additional video resources for some topics are listed in this *Companion Guide*.

Essay Assignments can also be used as Topics for Classroom Discussion. **Suggested Readings** in the text are resources for these assignments.

In writing essays students can come to understand the extent of their mathematical knowledge. This activity also familiarizes them with the language of mathematics—improving dialogue in the classroom.

Problems can also be used in the classroom as group activities (see this *Companion Guide*, Part iv, *Sample Group Activities*). Group activities encourage students to assume a more active role in the classroom, helping them to see themselves and their classmates (rather than only the instructor) as resources for learning. **Suggested Readings** in the text are resources for these assignments.

Computer Problems: In a course such as we are laying out, computer problems can serve to emphasize a number of important points. Among them are: (1) the verification of mathematical statements and identities; (2) the discovery of new facts through computer experimentation and induction (*not* mathematical induction); (3) the great successes and occasional pitfalls of the computation process; (4) the limitations imposed by the digital language as opposed to the richer existential language of "full" mathematics; (5) the extent to which one needs mathematical knowledge and expertise beyond what is built into commercial mathematical software; (6) the appreciation of the computational infrastructure of our civilization, an infrastructure that is often hidden from view.

IN THE COMPANION GUIDE

Expository Research Papers: Suggestions are given (see Part VI, *Suggestions for Expository Research Papers*) for expository research paper assignments. However, students should also be encouraged to choose their own topics. It is important to provide specific tasks for the students regarding the writing of a research paper and to give them explicit deadlines. These tasks include writing a one-page essay describing the topic they select and what they expect to learn and demonstrate in their paper; submitting a bibliography for the paper; writing an outline; submitting a first draft (see Part III, *Sample Syllabus*).

Topics for Classroom Discussion are useful for general class discussion or as a source of group activities. **Suggested Readings** in the text are resources for these discussions and additional video resources for some topics are listed in this *Companion Guide*. The emphasis for these topics is on *discussion* more than *lecture*. Relinquishing the lecture as the primary mode of instruction, taking the opportunity to hear the student in the classroom, helps to foster an environment in which instructor and students converse to form a community of learners. Conducting class meetings as open forums to discover student interest in selected topics, to motivate and explore objectives for topics, and to discuss possible directions that these topics suggest, helps students recognize their responsibility in the learning process.

Projects are connected sequences of explorations that start with very familiar, "easy" material and gradually lead the student to new discoveries and a glimpse at broader vistas. A project can be assigned as individual homework, worth perhaps two weeks each. Better, it can be carried out in small groups which periodically report to the class as a whole to compare notes. This reporting and comparing leads to work on the project by the class as a whole, with occasional hints and suggestions by the instructor. Such a class session will re-energize the individual students or groups to go further.

Projects can be shortened by omitting the last parts, or enlarged by instructor or students coming up with new directions to pursue. These projects are also models to help the

Special Features

instructor make up projects in line with her background and interests.

Tutorials: Two of the additions to this Companion Guide have been labelled "Tutorials." One is about differential and integral calculus, and the other is about complex numbers. These two topics are essential in any survey of mathematics. We suggest that the instructor provide the class with photocopies of this material.

Each of these sections could have been part of the original text. They are more text-bookish than the rest of the text, since they are straightforward presentations of classical mathematics. But we still strive for a respectable level of literary style, and take every opportunity to tell about historical background and philosophical controversies. The tutorial on calculus is a *tour de force*. In only fifteen pages it explains the guts of the usual one-semester course, including applications and problems.

Pictures and diagrams are essential when teaching calculus. Our artwork was done by Caroline Smith of the University of New Mexico, to whom we are most grateful.

Part II
Chapter Guidelines

Chapter 1
The Mathematical Landscape

What is Mathematics? Where is Mathematics? The Mathematical Community. Tools of the Trade. How Much Mathematics is Now Known? Ulam's Dilemma. How Much Mathematics Can There Be?

Topics for Classroom Discussion

1. An alien has landed. She asks you what mathematics is. How do you answer? What do mathematicians do? Does mathematics change?

2. What is Ulam's dilemma? If you were Ulam, and a reporter for *Newsweek* was interviewing you regarding this dilemma, how would you explain it? Is there anything that can be done about it? Will the *Information Superhighway* help?

3. Can mathematics establish truth? Plato thought so. Eric Temple Bell thought not. What do you think? How are proof and truth related? See, for example, "The Concept of Mathematical Truth" by Gian-Carlo Rota in *Essays in Humanistic Mathematics* (Washington, D.C., Mathematical Association of America, 1993).

4. Some mathematicians have described the process of mathematical research as a kind of "playing around." Discuss.

5. π is the ratio of a circle's circumference to its diameter. It cannot be constructed with a straightedge and a compass. π is irrational. It is a transcendental number. π shows up in number theory, in geometry, in probability. Was π invented or discovered by mathematicians? See, for example, "π and e" by E. C. Titchmarsh in *Mathematics: People, Problems, Results* edited by D. Campbell and J. Higgens (Belmont, CA: Wordsworth International, 1984).

Chapter Guidelines

6. Investigate the number *e*. What kind of number is it? Where do we find it in mathematics? Was it invented or discovered by mathematicians? A good reference is *e – The Story of a Number,* by Eli Maor (Princeton: Princeton University Press, 1994).

7. Find reasons for thinking that any brief definition of mathematics must be inadequate.

8. Defend the view that computer science is part of mathematics.

9. Could there be such a thing as "unconscious mathematics" which need not be symbolized in any way, but which leads to certain consequences?

10. How much music is there? How much literature is there? In these instances, how could you make an acceptable definition of "how much"? How could you go about implementing your definition? What could you do with an answer?

Chapter 2
Varieties of Mathematical Experience

The Ideal Mathematician. The Individual and the Culture. The Current Individual and Collective Consciousness. A Physicist Looks as Mathematics. I. R. Shafarevitch and the New Neoplatonism. Unorthodoxies

I. Discussion Topic: Proof

1. In "The Ideal Mathematician" the student asks the ideal mathematician about proof. What is *your* conception of proof? How would you answer the following questions:

Ch. 2: Varieties of Mathematical Experience

    a. What is the role of proof in mathematics?
    b. Why do mathematicians prove theorems?
    c. What does it mean for a mathematical statement to be considered true?

    2. Christian Goldbach (1690–1764) conjectured that every even number greater than two is the sum of two odd primes. Do you believe this? Why? Pick ten even numbers and see if they each are the sum of two odd primes.

    3. What role do intuition and evidence play in proof? What is the difference between conjecture and a proof?

    4. Kurt Gödel (1906–1978) showed that not every true statement is provable in mathematics. Is every provable statement true in mathematics?

II. Discussion Topic: The Many Roads to Proof

    1. Is there only one type of proof? If someone asked you to describe what a proof is and what it does, what would you say?

    2. The MATHEMATICS! videotape "The Theorem of Pythagoras" (Pasadena: California Institute of Technology, 1988) demonstrates several animated dissection proofs of the Pythagorean theorem. Students can be challenged to construct their own dissection proof or one they see in the videotape.

A nice follow-up to seeing this video is the geometric paradox one encounters in dissecting an 8-inch square into a 5×13-inch rectangle (see *Fibonacci Numbers* by N. N. Vorob'ev). It's very effective in demonstrating the difference between evidence and proof. Fibonacci numbers can be introduced here or later to revisit the paradox. (See page 35 of this *Companion Guide*.)

    3. Do you think there is proof in the following professional areas: medicine, physics, law, religion?

    4. How would you *prove* that $123 \times 587 = 72201$?

III. Tutorial: Calculus

This is a concise presentation of the main ideas of calculus. Because the topic is so important, we make some unavoidable concessions to the standard textbook format.

Chapter Guidelines

Calculus is the heart of modern mathematics, since Newton. It's the part of mathematics most important in science and technology, the part engineers must know.

It grows out of two main problems which at first seem unrelated.

The central discovery of calculus is that these problems are opposites or inverses.

The first main problem in calculus is speed. How fast is something changing? The solution of this problem is "the differential calculus." The second main problem is area. How big is the inside of some curved region? The solution of this problem is "the integral calculus."

First we'll talk about speed. It's easy to find the speed if it's constant. Just divide the distance traveled by the time elapsed. Speed = Distance/Time.

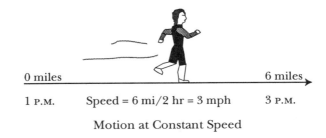

Motion at Constant Speed

But in real motion, speed isn't constant. You start your car at speed zero. You go faster till you get to the speed limit.

Car goes from
Albuquerque to Santa Fe, variable speed

Motion at Variable Speed

Then to stop, you slow back down to zero. Your speed changes from instant to instant. What is your speed at some particular instant?

Here's another practical example. Nick, our math professor, falls off the First International Unpaid Debts Building in Miami. How fast is he falling? In school we learned that in a vacuum, under the acceleration of gravity, a body falls $16t^2$ feet in t seconds. How fast is he falling after 2 seconds, ignoring air resistance?

Nick falling off the F.I.U.D. Building.

In the time interval between 2 seconds and 2.1 seconds—time lapse of 0.1 second—the distance Nick falls is the difference between how far he had fallen after 2 seconds and how far he had fallen after 2.1 seconds.

$$16(2.1)^2 - 16(2^2) \text{ feet} = 6.56 \text{ feet}.$$

Dividing distance by time (0.1 second), his average speed over that tenth of a second is 65.6 ft/sec. That's the constant speed that would carry him the same distance *in the same time as did his actual fall at the accelerated speed.*

Exercise. Repeat the calculation with a time lapse of 0.01 second. (You'll get an average speed of 64.16 ft/sec, between time 2 seconds and time 2.01 seconds.)

Chapter Guidelines

Do it still again, with a *tiny* time lapse, 0.001 seconds. (Nick's average velocity over this time period is 64.01 ft/sec.)

We don't want an *average* speed. We want the *exact* speed at time 2! That means a time lapse of *zero*. The formula Speed = Distance/Time breaks down, because division by zero is meaningless. However, *without setting the time lapse equal to* 0 you've crept closer and closer to 0. You used lapses of 0.1, 0.01, 0.001, and found speeds of 65.6, 64.16, 64.016.

NOW!! A giant conceptual leap! If the average speeds approach a limit as the time lapse approaches zero, we declare, as a *definition*, that this limit *is* the instantaneous speed!! In this example, the limit is 64 ft/sec when $t = 2$. It makes sense! It works! That's what we mean by instantaneous rate of speed or velocity.

NEWTON

LEIBNIZ

BERKELEY

ROBINSON

Ch. 2: Varieties of Mathematical Experience

This notion of rate as a *limit* took hundreds of years to formulate. Medieval and Renaissance mathematicians calculated a few rates of change without defining mathematically what they wanted. The founders of calculus, Isaac Newton and Gottfried Leibniz, enjoyed a bitter quarrel about priority in the discovery.

Leibniz' explanation of differentiation was not quite the same as Newton's. Leibniz used an *infinitesimal* increment—a number bigger than zero, yet smaller than any ordinary number.

Then George Berkeley, an empiricist philosopher and Anglican bishop, showed that the reasoning of both Newton and Leibniz was illegitimate. The small increment is sometimes defined to be not zero, sometimes to be zero. This is a contradiction! It was hundreds of years before an answer to Berkeley was found. But meanwhile mathematicians went on with the calculus anyway.

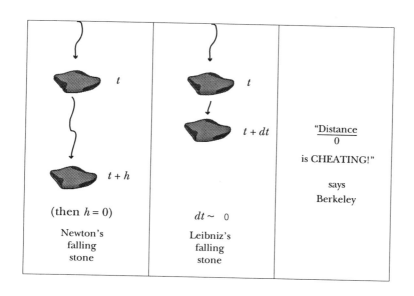

For centuries, people doubted whether infinitesimals make sense. In the 1960s an American logician, Abraham Robinson, used methods from modern logic to make the infinitesimal respectable.

Chapter Guidelines

Exercise. Make a graph of this falling body function: distance = time squared or $d = t^2$. (We dropped the 16 to simplify your graphing and our calculating.) This is a quadratic function. Its graph is a parabola. You've studied parabolas, but this calculation is different. Mark the points (2, 4) and (2.1, 4.41) on the parabola. The second is above and right of the first. Draw a straight line (called the "secant") between the two. What's the slope of this line? ("Rise over run.") Rise = $2.1^2 - 2^2 = 0.41$. Run = $2.1 - 2 = 0.1$. Slope = $0.41/0.1 = 4.1$, *which we just found is also the average velocity* (allowing for the factor of 16 which we took out). The average rate of change of distance as a function of time is identical to the slope of its graph! Again, replace 0.1 by 0.01 and 0.001. The corresponding marks on the graph are creeping closer and closer to (2, 4). The slopes of the secants are exactly the numbers you found to approximate the instantaneous rate of change. "In the limit," as the two points approach closer and closer, and the denominator approaches zero, the secant becomes a tangent and its slope becomes the instantaneous speed, called the *derivative*.

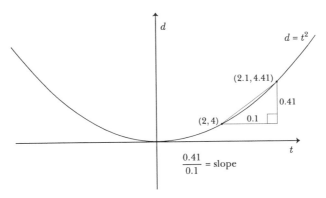

Differentiating x^2 is the same as finding its shape. (This graph is called a parabola.)

The process of calculating the derivative (the speed) is called differentiation. Simple functions usually have simple derivatives. The derivative of t^n is nt^{n-1}. (n is any number, integer or fraction, positive or negative.) The derivative of the natural logarithm of t is $1/t$. The derivative of e^{at} is ae^{at}. The

Ch. 2: Varieties of Mathematical Experience

derivative of $\sin t$ is $\cos t$; of $\cos t$, $-\sin t$. These formulas are always derived in first semester calculus.

Exercise. In a way similar to how you found the rate of change of $f(t) = t^2$ at $t = 2$, find the rate of change of that function at an arbitrary time t. Do the same for the cubic $f(t) = t^3$. Check your answer with the formula in the previous paragraph for t^n.

Now you're ready for the second main problem of calculus, integration—finding the area inside a curve. To solve it, strangely enough, we'll talk about another, quite different-sounding problem. Given the velocity of a moving body—say a car driving down the highway—can we calculate the total distance traveled, at any instant of the trip? This is the opposite of the problem we analyzed above. There we were given the distance and found the velocity.

Start with the simplest case—constant velocity. Suppose that from 2 PM to 3 PM you're driving at a steady 50 miles per hour. How far do you go in that hour? In half an hour? At any time t between 2 PM and 3 PM?

Of course you can answer this—in one hour, 50 miles. In half an hour, 25 miles. In t hours, $50t$ miles—where t can be a fraction.

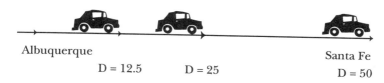

Driving 50 miles at a constant speed of 50 m.p.h.

The graph of these facts is simple. Time elapsed is measured on the horizontal time axis. The graph of the constant velocity 50 is a horizontal line 50 units above the time axis. We compute distance by multiplying speed times time—that is, height of the velocity line times length of the time axis from start to finish. The product of these horizontal and vertical lengths equals the area of the rectangle they enclose. Distance is represented graphically as area!

The real problem comes when you vary the speed of your car. Then the graph of $v(t)$, velocity as a function of time, is

Chapter Guidelines

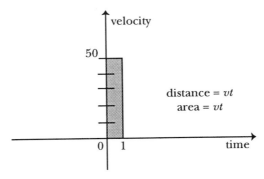

On a time–velocity graph, distance = area.

a curve, not a horizontal line. How can we find the distance traveled now? Since we know how to do it in the case of constant speed (horizontal graph), *replace the curved graph by a piecewise horizontal graph.*

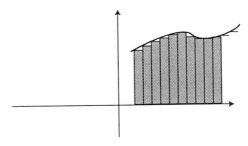

Variable velocity is approximated by piecewise-constant velocity.

In other words, instead of a speed varying smoothly, make the speed constant for a second, then a different constant for the next second. The distance traveled in each second is the speed in miles per second, and is shown in the graph as the area of a skinny vertical rectangle, of width one second. The areas of the little rectangles under the velocity curve add up to something close to the total distance, and also to the total area under the curve. So we see that in the case of varying speed, as in the case of constant speed, *the distance traveled is equal to the area under the velocity curve.*

To summarize: To a *distance* function $d(t)$ is associated a velocity function $v(t)$, the derivative of $d(t)$. To $v(t)$ in turn

Ch. 2: Varieties of Mathematical Experience

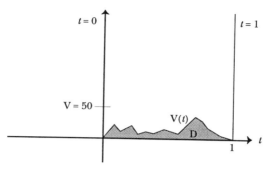

Distance = area under *any* velocity graph (variable speed).

is associated an *area* function $A(t)$, the area under the graph of $v(t)$ up to the vertical line t. *The area $A(t)$ is equal to the distance $d(t)$, the antiderivative of $v(t)$.* The area $A(t)$ under the graph of $v(t)$ is called the "integral" of $v(t)$. The function $d(t)$, from which $v(t)$ was obtained by differentiation, is the antiderivative of $v(t)$. Finding $A(t)$ is called "integrating" $v(t)$. We have just proved the "Fundamental Theorem of Calculus": The area function of v (the integral of v) is equal to the antiderivative of v:

$$A(t) = d(t).$$

We have been thinking of $v(t)$ as a "velocity function." But any function can be interpreted as a velocity function! So the Fundamental Theorem says: The integral of the derivative of any function is the function itself (except possibly for an additive constant).

Computing the derivative directly from its definition is often easy; computing the integral directly from its definition can be hard. The Fundamental Theorem let's us do the hard part by doing the easy part: make a dictionary of differentiation formulas. If in your collection you find a function $w(t)$ whose derivative is $v(t)$, then $w(t)$ is the integral of v!

Let's do an example simple enough that we can get the area by a direct integration. We'll take $v(t) = t$. Not a very realistic velocity function—you'd soon get a speeding ticket! But never mind, this is just theory.

Let's call the region whose area we want to evaluate D. The lower boundary of D is a portion of the positive x-axis; the upper boundary is part of the graph of $v(t) = t$. This graph

21

Chapter Guidelines

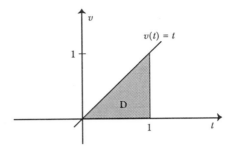

Distance = area under speed graph if speed = time.

is a line through the origin with slope 1. The left boundary, $t = 0$, is the point where upper and lower boundaries intersect. The right boundary is a vertical segment, $0 < y < t$. You see immediately that D is an isosceles right triangle with sides of length t inches.

Using the triangle area formula (one-half base times altitude), you find the area $A = t^2/2$ square inches. Next you'll find the area by calculus. Why do it twice? Because the triangle area formula only works for triangles and polygons. The calculus works for curved areas too.

What you do is cut up the triangle with vertical lines 0.01 inches apart. This makes a lot of long, skinny pieces, almost rectangles, with a tiny little triangle at the top of each rectan-

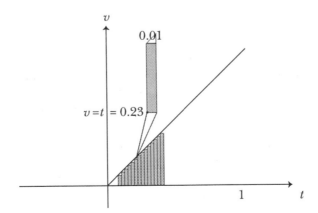

Area $\sim$ sum of skinny rectangles if velocity = time.

Ch. 2: Varieties of Mathematical Experience

gle. What if you ignore those tiny little triangles? Each rectangle is 0.01 inches wide. How high? The upper boundary of each rectangle is part of the line $y = t$. The point of the graph above $x = 0.23$, for example, has y-coordinate $y = x = 0.23$. That's the height of the rectangle inside the 23rd piece. The area of any rectangle is height times width, so the area of this rectangle is 0.01 times 0.23 = 0.0023 square inches.

To get the whole area, add the areas of all these rectangles. Just as the 23rd has area 0.0023 square inches, the 38th will have area 0.0038 square inches, and so on. Adding them all up, and factoring out 0.0001, you get for your approximation to the area of the whole triangle

$$0.0001 \times (1 + 2 + 3 + \cdots + 100 + \cdots).$$

How far should the sum go? Well, how many rectangles are there? You have a base t inches long, and you cut it into pieces 0.01 inch = 1/100 inch wide. So there are $100t$ rectangles. The last term of the sum in parentheses is $100t$.

This is a nice puzzle: $1 + 2 + 3 + \cdots + 100t = ?$

A lovely trick does the job. It was discovered by the famous mathematician Karl Friedrich Gauss in school in the first grade. A mean teacher set the class to add all the numbers up to 100.

Karl noticed he could write the sum twice—once forward, once backward. The number in the first sum plus its neighbor below in the second sum always add up to 101. He had 100 such pairs. So the two sums together equal 100 times 101. The single separate sum would be half of that: 5,050.

With young Karl as model, you'll find that the sum of the numbers up to $100t$ is $50t$ times $(100t+1)$, which is $5000t^2 + 50t$. Look back and remember, this had to be multiplied by 0.0001. Since $(5000)(0.0001) = 0.5$, we get $0.5t^2 + 50t(0.0001)$. The first term, $0.5 t^2$, is the exact answer, as we already know from geometry. The second term is "the error." It's the total area of those little triangles we neglected. It isn't such a big error—$t/200$. If we replace 0.01 by 0.001, the error will be $50 t$ divided by 10^6 square inches, instead of 10^4.

But why stop at 0.001 or 0.0001? Your calculations still work, even if the rectangles are so thin you can't see them. You're taking the width of the little rectangle as a "parameter" which

Chapter Guidelines

$$1 + 2 + 3 + \ldots + 99 + 100 = S$$
$$100 + 99 + 98 + \ldots + 2 + 1 = S$$
$$\overline{101 + 101 + 101 \ldots + 101 + 101 = 2S}$$

Young Gauss with double sum, thinking "Aha!"

you're "sending to zero." If you set it really close to zero (but not equal to zero) you get an error so small it just doesn't matter.

(It wouldn't do to set the thickness of each little piece *equal to* zero. Then each little rectangle would have area zero, and they'd all add up to *zero*.)

Calculating area by adding many tiny rectangles is called "integration." As in calculating instantaneous speed in the differential calculus, the method works, it makes sense, so the area inside a curve is *defined* to be the limit of the areas of inscribed or circumscribed polygons. There can't be a *proof* that the limit equals the area, because we have no other definition of area, except that limit!

What you have accomplished is not just a complicated way to measure triangles. It works for about any area that comes along. Problems on arc length, volume, probability, mass, elec-

Ch. 2: Varieties of Mathematical Experience

trical capacity, work, inertia, linear and angular momentum, all lead to integrations such as you just did.

Now let's see what happens if we do our calculus operations in the opposite order—first integrate, then differentiate. Before, we started with a distance function, differentiated to get a velocity function, then integrated that and got back our original distance function. Now we start with a velocity function, integrate it to get an area and distance function, then differentiate that, to get—what do you suppose?

Look back at the last example, but now let the right side of the triangle be movable. In other words, t, the distance from the origin, is variable. The right boundary is a stick you slide right and left. To the right, area increases. To the left, area decreases.

You're going to calculate the rate of change of A (the area of D) as the stick moves to the right. That is, you will differentiate the integral.

D need no longer be a triangle. It's still bounded below by the t-axis, on the left by the y-axis, and on the right by the vertical line at t, but its upper boundary is the graph of any function $v(t)$ you like.

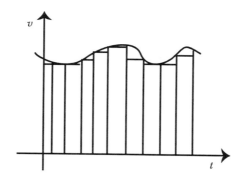

Area $\sim$ sum of skinny rectangles for any graph.

To differentiate—find a rate of change—increase the independent variable t by a tiny little bit h see how much your function increases from $A(t)$ by $A(t+h)$, and then divide this increment of area by h. This quotient is the average increase over the interval from t to $t+h$. Since h is small, the numerator

Chapter Guidelines

and denominator are both close to zero. Their ratio is close to a limit, which limit you defined as the rate of change of $A(t)$.

In applying the definition of rate of change to the area function $A(t)$, you're working with *two different pictures*. The integration picture computes the area of D, the region under $v(t)$, by cutting up D with many close vertical lines. The differentiation picture computes the rate of change of any function by drawing the secant through two nearby points on its graph. We're applying the differentiation picture to the integration picture, or, if you like, plugging the integration picture into the differentiation picture.

You're differentiating $A(t)$, the area of D bounded by the vertical line at t. Take your definition of the area A from the integration picture. D is the region under the graph of $v(t)$. What happens to D and its area A if its right side moves a bit further right? The region is enlarged by a tiny additional piece, which differs from a rectangle only in a super-tiny bit at the top. Its width is h, the amount of increase of t. Its height is the height of the upper boundary of D, which is the graph of $v(t)$. So the height of the little added rectangle is $v(t)$, and its area is $hv(t)$. This $hv(t)$ is the increment of $A(t)$. The derivative of $A(t)$ is the increment $hv(t)$ divided by h.

That's $hv(t)/h = v(t)$!

We have just shown that the derivative of A is v. But A is the integral of v. The derivative of the integral of v equals the integral of the derivative of v equals—v. Symbolically,

$$D: \quad A \longrightarrow v$$
$$I: \quad v \longrightarrow A.$$

We have shown that in either order, differentiation and integration reverse each other. This is called "the Fundamental Theorem of Calculus."

The Fundamental Theorem of Calculus

$$IDs = DIs = s$$

(Rate of change of area under velocity graph)

= (rate of change of distance)

= (height of velocity graph)

Ch. 2: Varieties of Mathematical Experience

The Fundamental Theorem gives a powerful method of computing areas. Suppose we want to know the area A under the parabola $y = 3x^2$, between $x = 0$ and $x = 3$.

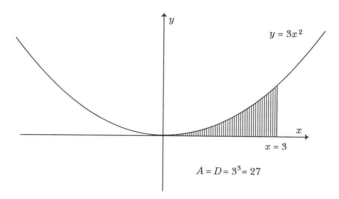

An area under a curve, by the Fundamental Theorem.

According to your work a few paragraphs above, this function is the derivative of x^3. Therefore, by the Fundamental Theorem, its area function $A(x)$ is x^3. Between 0 and 2, the area under $3x^2$ is therefore $2^3 - 0^3 = 8$ square inches.

Exercise. Use the Fundamental Theorem to compute the areas under the curves $y = x^2$, $y = x^3$, $y = x^4$, $0 < x < 1$.

The most important use of calculus is in solving differential equations. These involve an unknown function and its first and second derivatives. In case the unknown function is a distance, the first derivative is the velocity, and the second derivative, the rate of change of the velocity, or the "acceleration." The fundamental law of mechanics, Newton's third law, says $f = ma$—force equals mass times acceleration. Often the force is given by some fundamental principle governing the motion under study. Since acceleration is a second derivative of position, Newton's law is a second-order differential equation. To find out how a body moves under the influence of a force, we have to solve this differential equation.

In the case of the planets and the sun, the force is gravity, which is directly proportional to the masses of the attracting bodies, and inversely proportional to the square of their

Chapter Guidelines

distance. In the case of only two bodies, such as Earth and the Sun, the differential equation can be solved. By doing so, Newton proved that the three laws of Kepler (elliptic orbits; position vector covering equal areas in equal times; and the length of year proportional to the $4/3$-power of the orbits) are equivalent to the law of gravity and the third law of motion. But this calculation requires more technique than we have mastered here.

This triumph of Newtonian calculus and physics is based on ignoring the mutual attractions of the planets. If we think of Mars, Earth, and the Sun as a system of three bodies, none of whose mutual interaction should be ignored, we have the stubbornly intractable three-body problem, which has been a source of frustration for 300 years.

In order to get a glimpse of how differential equations solve problems of motion, consider a ball thrown into the air in a room with an M-foot ceiling. I want the ball to just barely touch the ceiling. This depends on the velocity V with which I toss the ball up. Given M, can I determine V?

How hard to throw to barely touch the ceiling?

As in all elementary treatments, we ignore air resistance. The ball does not encounter any friction. The only force is gravity, which creates a downward acceleration of 32 feet per second^2. The initial height of the ball is zero, since we measure from ground level. Now Newton's third law is

$$\text{Mass} \times \text{Acceleration} = -32 \times \text{Mass}.$$

Ch. 2: Varieties of Mathematical Experience

(The minus sign because acceleration due to gravity is downward, decreasing $h(t)$.) This equation can be solved by two integrations. First divide through by the mass. Since acceleration is the derivative of velocity, and since -32 is the derivative of $-32t + K$, where K is any constant, the integration gives

$$\text{Velocity} = -32t + K.$$

We determine K by setting $t = 0$. The equation becomes "initial velocity $= 0 + K$," so $K = V$, the initial velocity, which we ultimately want to determine as a function of M. Velocity $= -32t + V$.

Since velocity is the derivative of height, and $-32t + V$ is the derivative of $-16t^2 + Vt + L$, where L again is an arbitrary constant, integration gives

$$h = -16t^2 + Vt + L.$$

Again set $t = 0$. Since $h(0)$, the initial height is 0, we get $L = 0$. So we have derived a formula for height as a function of time: $h(t) = -16t^2 + Vt$.

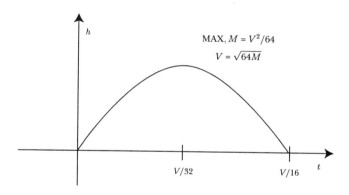

Height of the ball as a function of time: up, then down.

Finally, what is M, the maximum of h? When the ball reaches its maximum M, it ceases its rise and is about to fall. At that instant its velocity is neither positive (up) nor negative (down). It's zero. But we have found a formula, Velocity $= -32t + V$. Therefore, when Velocity $= 0$, $V = 32t$, $t = V/32$. So the time when the ball is at its highest is $V/32$. When $t = V/32$, our

Chapter Guidelines

formula for $h(t)$ simplifies to $h(t) = V^2/64$, so $V^2/64 = M$. There's a tiny simplification for you to check.

Solving for V, we find that V is the square root of $64M$. For instance, if $M = 100$ feet, V should be 80 feet per second.

We computed this figure for ball-playing here on Earth, where g, the acceleration due to gravity, is 32 feet per second per second. What if we visit the Moon? Or Mars? We'd have to replace 32 by the gravitational constant there. On the moon g is much smaller, so we'd need a smaller V for a given M.

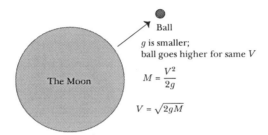

Weaker gravity, higher throw.

We could repeat the whole calculation, starting with the Moon's g instead of 32. Or we can just look at our Earthly answer—V = the square root of $64M$—and make the obvious guess for Mars or the Moon—V = the square root of $2gM$.

Finally, let's go back to Nick, the math professor who fell off the First International Unpaid Debts Building in Miami.

The fire department, thank goodness, is at the foot of the building, life net ready, to catch Nick before he gets hurt. But is the net strong enough? Let's say Nick weighs 250 pounds, and the building is 2600 feet tall. How hard is he going to hit that life net? In other words, how fast will he be going when he reaches the ground?

Like the planets and the ball, he is subject to Newton's third law, with gravitational force -32 feet per second per second. In the equation "acceleration = -32," we recognize that both sides are derivatives, or rates of change. Acceleration is the rate of change of velocity, and -32 is the rate of change or derivative of $-32t$ plus an arbitrary constant A. So, velocity $= -32t + A$. What's A? Set $t = 0$. The velocity equation now

Ch. 2: Varieties of Mathematical Experience

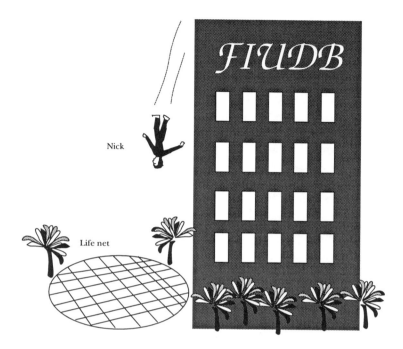

Will Nick be saved?

reads: initial velocity = A. What was his initial velocity? When he had just slipped off the roof, he hadn't started to fall yet. His velocity was zero. So $A = 0$. Again, in the equation "velocity = $-32t$," we recognize that both sides are rates of change. Velocity is the rate of change of position $h(t)$, and $-32t$ is the derivative of $-16t^2$ plus an arbitrary constant B. At the time $t = 0$, Nick had not yet fallen any distance, so $B = 0$, and $h(t) = -16t^2$.

Nick will arrive at ground level when his distance fallen, $16t^2$, equals 1600, the height of the building. That's when $t^2 = 100$, or $t = 10$. So his velocity on hitting the life net is 32 times 10 = 320 feet/second. Multiplying by his weight, we find that the life net must withstand a momentum of 250 times 320, or 80,000 foot-pounds per second.

There's no way to get this information except by calculus and differential equations.

Chapter 3
Outer Issues

Why Mathematics Works: A Conventionalist Answer. Utility. Number Mysticism. On the Utility of Mathematics to Other Scientific or Technological Fields. Pure vs. Applied Mathematics. On the Utility of Mathematics to Mathematics. Mathematical Models. Underneath the Fig Leaf

I. Discussion Topic: Mathematics and Art

1. What are the relations of mathematics to art? Is computer "art" really art? In what ways have artists used mathematics, and/or computers?

2. In *Descartes' Dream* by P. J. Davis and R. Hersh (Boston: Harcourt, Brace, Jovanovich, 1986), artists are called "unconscious mathematicians." Discuss the different kinds of mathematics which artists discover, use, or create.

3. The Golden Rectangle and the Golden Ratio in art. See, for example, *For All Practical Purposes*, edited by L. A. Steen (New York: W.H. Freeman, 1987).

4. Is there beauty in geometric figures? Is there beauty in reasoning?

II. Discussion Topic: Is Mathematics Created or Discovered?

1. Cantor's continuum hypothesis and the reason why constructivists reject it.

2. How does the mathematics-by-fiat view differ from the Platonic view of mathematics that believes the universe imposes mathematics on humanity?

3. Is π an example of creation or discovery in mathematics? Geometrically we describe π as the ratio of a circle's circumference to its diameter. It was not determined until 1767 that π

Ch. 3: Outer Issues

is irrational (Johann Lambert, 1728–1777). More than a century later Ferdinand Lindemann (1852–1939) proved it could not be the solution of any polynomial equation with integer coefficients. This discovery showed that the squaring of the circle is impossible. So the route from geometry to number theory comes full circle?

4. The Fibonacci numbers appear in pineapples and pine cones. Many patterns of biological growth can be described in terms of this sequence of numbers. How would the Platonist, the formalist, and the constructivist each explain this phenomenon? See, for example, *For All Practical Purposes*, edited by L. A. Steen (New York: W.H. Freeman, 1987).

Videotape Resources

Apostol, Tom: "The Story of π" (MATHEMATICS!: Pasadena: California Institute of Technology, 1988).

Freeman, W.: "On Shape and Size" (*For All Practical Purposes* (New York: W.H. Freeman, 1988) (30-minute videotape). The introduction includes a discussion of Fibonacci numbers and the golden ratio.

III. Discussion Topic: Mathematical Models

1. The traveling salesman problem (see V. K. Balakrishnan's *Introductory Discrete Mathematics*, Englewood Cliffs: Prentice Hall, 1991).

2. Leonardo of Pisa (ca. 1202) developed the Fibonacci sequence in trying to answer the following question:

> If someone places a pair of rabbits in a certain place enclosed on all sides by a wall, how many pairs of rabbits will be born there in the course of one year, it being assumed that every month a pair of rabbits produces another pair, and that rabbits begin to bear young two months after their own birth?

Describe how the Fibonacci sequence answers the question.

3. Consider the current civil calendar. Compare and contrast it with the Moslem, Hebrew, Chinese, Greek Orthodox, and "World" calendars.

4. Investigate the following models of the planetary system: Ptolemaic, Copernican, Keplerian, and Newtonian.

Chapter Guidelines

5. Read up on the history of ballistics and contrast the pre-Galilean and the post-Galilean theories of the trajectory of a missile.

6. Toss one die. Toss two dice. How would you assign the probabilities? How would you justify your assignment? How would you determine whether the dice were loaded?

7. Investigate models of decorative patterns. Discuss the group-theoretical model. See *Symmetries of Culture* by Dorothy K. Washburn and Donald W. Crowe (Seattle: University of Washington Press, 1988). If you went into a paint and wallpaper store, how would wallpapers be classified? Why?

8. Discuss cryptographic models. There are many ways of turning "plain text" into coded text. What are the pluses and minuses? See, for example, *Elementary Cryptanalysis* by A. Sinkov (Washington, D.C.: Mathematical Association of America, 1966).

9. Examine measures of intelligence. Is IQ an adequate mathematical model? Should IQ really be vector valued (i.e., many components to intelligence)? See *Frames of Mind: The Theory of Multiple Intelligences* by Howard Gardner (New York: Basic Books, 1983).

10. Discuss the mathematization of professional baseball. How do you mathematize baseball strategy? How do you measure the value of a player? Check out SABERMATRICS and the Society of American Baseball Research. Also see *Men at Work: The Craft of Baseball* by George F. Will (New York: Macmillan, 1990).

Videotape Resources

Freeman, W.: *For All Practical Purposes* (New York: W.H. Freeman, 1988). 30-minute videotapes:

Management Science—Overview: Apollo 11; Optimization: airline scheduling, limited constraints and resources, Bell Labs; Disney World: queueing; Avis: transportation; Street Smarts: street networks; Trains, Planes, and Critical Paths. Topics include the traveling salesman problem, algorithm for finding the cheapest tour, combinatorial explosion, nearest neighbor algorithm, telecommunications networks, minimum cost, spanning trees, Kruskal's algorithm, critical path problems; Scheduling Problems: "Juggling Machines." Topics include math modeling, machine scheduling, heuristic algorithms,

Ch. 3: Outer Issues

list-processing algorithms, bin packing; Linear Programming: "Juicy Problems." Topics include constraints, optimization, feasible points, feasible sets, corner principle, production policy, simplex method.

Social Choice—Election Theory: "The Impossible Dream": mathematics of decision making, outcomes, game theory, fair division; Weighted Voting: "More Equal Than Others": weighted voting systems, Banzhaf power index, apportionment; Zero Sum Games: games of conflict, problems of fair division; Prisoner's Dilemma: game theory, games of partial conflict, cooperation and defection.

IV. Discussion Topic: Number Mysticism

1. Explore the figurate numbers of the Pythagoreans and discover some number facts arising from them. See "Pythagorean Arithmetic" by R. Honsberger in *Ingenuity in Mathematics* (Washington, D.C.: Mathematical Association of America, 1970).

2. Develop some conjectures about the Fibonacci numbers.

Videotape Resources

Apostol, Tom: "The Theorem of Pythagoras" (MATHEMATICS!: Pasadena: California Institute of Technology, 1988).

V. Discussion Topic: Utility of Mathematics to Other Scientific or Technological Fields, and Vice-Versa

1. Many different answers have been given to the question of why mathematics is useful in describing the physical workings of the universe. Find out what some of these answers have been. Which one appeals to you most? Why?

2. Examine what role the Pythagorean theorem plays in Einstein's theory of relativity. See K. O. Friedrich's book *From Pythagoras to Einstein*, published by the Mathematical Association of America, 1965).

Films

Molecular Spectroscopy, by Reid H. Ray, Film Industries, Inc., 1962; Chemical Education Material Study; Dr. Bryce Crawford and Dr. John Overland, University of Minnesota. This film includes applications of symmetries in chemistry. It discusses how polyatomic molecules vibrate in symmetrical or asymmetrical fashion. The symmetry of the molecules can be correlated to the modes of vibration. Rotational motions of the molecules are explained.

Chapter Guidelines

VI. Discussion Topic: Pure vs. Applied Mathematics

1. Archimedes was not satisfied with certain arguments or proofs based upon mechanics. Why? See *The Works of Archimedes*, T. L. Heath, ed. (New York: Dover, 1953), Supplement, pp. 7–14. Also see *Theology and the Scientific Imagination*, Amos Funkenstein (Princeton: Princeton University Press, 1988, Chapter V).

For questions 2, 3, and 4, see for example *Mathematics: People, Problems, Results* edited by D. Campbell and J. Higgens (Belmont, CA: Wordwsorth International, 1984).

2. Does the pursuit of applied mathematics, as opposed to pure mathematics, require a different type of personality?

3. Which is a harder subject: pure mathematics or applied mathematics? Support your conclusions.

4. What kind of employment can a pure mathematician find? An applied mathematician?

VII. Discussion Topic: On the Utility of Mathematics to Mathematics

1. Discuss how mathematicians often use results from one branch of mathematics to prove results in another. Examples: the four-color theorem, Fermat's last theorem.

2. Recall the dissection proof of Chapter 2, Problem 2 in the text. Demonstrate how Fibonacci numbers can be used to describe the one-unit discrepancy when you dissect an 8-inch square and try to transform it into a 5×13-inch rectangle.

3. Investigate how Pythagorean triples can be used to generate Fibonacci numbers. See "Connections in mathematics: An introduction to Fibonacci via Pythagoras" by E.A. Marchisotto (*The Fibonacci Quarterly*, Vol. 31, No. 1, February 1993).

Chapter 4
Inner Issues

Pattern, Order, and Chaos. Abstraction, Generalization, Formulation. Proof. The Aesthetic Component. Mathematical Objects and Structures; Existence.

I. Discussion Topic: Pattern, Order, and Chaos

1. Give examples from life and from mathematics that illustrate order out of order; chaos out of chaos; order out of chaos; and chaos out of order.

2. Frieze patterns, i.e., one-dimensional patterns.

3. Two-dimensional patterns (see, for example, Washburn and Crowe, *Symmetries of Culture: Theory and Practice of Plane Pattern Analysis*, Seattle: University of Washington Press, 1988).

II. Discussion Topic: Proof and the Different Ways of Proving

1. Do you find Euclid's proof of the Pythagorean theorem convincing? Why or why not? Look up one of the proofs by dissection in E.S. Loomis' book, *The Pythagorean Proposition* (Washington, D.C.: National Council of Teachers of Mathematics, 1968). Is it more or less convincing than Euclid's?

2. Prosecutors in court claim to give proof of guilt. Is this the same thing as mathematical proof? Does it have anything in common with mathematical proof?

3. Logical theory says that if the assumptions are true and the proof is correct, the conclusions must be true, with no possible doubt whatever. Do you agree? Why or why not?

4. Mathematical induction: There is a castle with an infinite number of rooms. The king has the key to room #1. When he enters it, he finds the key to room #2. If in every room there is the key to the next room, will the king be able to enter all

Chapter Guidelines

the rooms of the castle? The analogy to this scenario is proof by mathematical induction.

5. A proof by contradiction establishes a certain statement by showing that its denial leads to a contradiction. Euclid's proof of the infinitude of primes is an example of such a proof. See *Journey Through Genius* by W. Dunham (New York: John Wiley & Sons, 1990).

6. The use of counterexample to disprove conjectures in mathematics.

III. Discussion Topic: Pascal's Triangle

1. Pascal's triangle let's you solve many problems involving counting and choice. For example: While driving home from work at rush hour you must pass through eight intersections controlled by stoplights. When you reach an intersection the light is either green or red. Use Pascal's triangle to solve these problems:

   a. In how many ways can you go through the intersections catching *at least* four green lights?

   b. In how many ways can you go through the intersections in which *all* the lights are green?

   c. In how many ways can you go through the intersections in which *no more than* two of the lights are green?

2. The sum of numbers lying above the nth rising diagonal of Pascal's triangle (including that diagonal) equals $u_{n+2} - 1$ (where u_{n+2} is the $(n+2)$th term of the Fibonacci sequence). Verify for several values of n. Now could you prove it? Look for other identities connecting Fibonacci numbers with binomial coefficients.

IV. Discussion Topic: The Aesthetic Component— Beauty in Mathematics

1. What objects or processes do you see as beautiful? Are there objects or processes in mathematics which can be so described?

2. Discuss, analyze, and criticize Edna St. Vincent Millay's poem, "Euclid Alone Has Looked on Beauty Bare."

Ch. 4: Inner Issues

3. Examine three different proofs of the formula

$$1 + 2 + 3 + \cdots + n = n(n+1)/2.$$

Which do you find most pleasing? Why?

4. Examine some frieze patterns (see *For All Practical Purposes*, edited by L. A. Steen [New York: W.H. Freeman, 1987]).

Do you have any preferences among them? Try to describe why some are more pleasing to you than others.

5. Of the mathematical curves you are familiar with, which do you consider the most interesting? Give reasons. Discuss Giuseppe Peano's space-filling curve (see *Mathematical Thought from Modern to Ancient Times* by M. Kline [New York: Oxford University Press, 1972], page 1018). What makes this interesting or surprising? Why is it called a curve?

V. Discussion Topic: Mathematical Objects and Structures—Existence—The Importance of Context in Mathematics

1. Does the fundamental theorem of arithmetic hold in Jonathan Swift's Kingdom of Laputa (*Gulliver's Travels*)? By "traveling" to Laputa (see Sherman Stein's *Mathematics, The Man-made Universe* [San Francisco: W. H. Freeman, 1963], page 32), students can compute Lagado primes and discover how a change of context can change results in mathematics.

2. Where does the Pythagorean theorem hold? Can an analog of the Pythagorean theorem be developed for three dimensions? Does the theorem ever hold on the surface of a sphere? What theorem holds on the surface of a sphere? (See, for example, "A New Look, Pythagoras" by C. Thornton in *The Mathematics Teacher*, vol. 74, no. 2, February 1981.)

3. React to the following statements:

   a. A tourist travels from St. Louis, Missouri to Sydney, Australia. She must cross the equator.

   b. On the basis of the information given in (a.), we can determine where the tourist has crossed the equator.

4. Let $M = 100^{37}$, and let $N = 37^{100}$. React to the following statements:

   a. We can find a prime number larger than $M + N$.

b. It's easy to find a prime number larger than $M + N$.

c. We can determine whether $MN+1$ is a prime number.

5. React to the following statement: You had an ancestor who was alive on September 1, 1066.

VI. Discussion Topic: Abstraction, Generalization, Formalism

1. Generalize this statement in two different ways: If the sides of a rectangle have length a and b, its area is ab.

2. Consider the following two statements and decide whether B is a generalization of A.

Let A be the statement: The medians of any triangle intersect in a single point.

Let B be the statement: The angle bisectors of any triangle intersect in a single point.

3. Referring to #2, can you find a statement that generalizes both A and B?

VII. Project Topic: Mathematical Relations and Equivalence Classes

Warning! This project is about "abstract relations." As often happens in mathematics, words have been stolen from plain English and given a "technical" meaning. The new meanings of "relation" and "equivalent" are related to, but not the same as the familiar plain English meanings.

We may be interested in the logical consequences of two objects being more than a mile apart, as is true, for example, of Boston and Austin. And we may abbreviate the statement "Austin and Boston are more than a mile apart" with some symbol—perhaps some letter of the alphabet—say R. Then we can write, for short, "Boston R Austin" or "Austin R Boston." We can even speak of the relation R as a thing in itself, and say "Boston and Austin are in relation R."

Since Albuquerque, Hersh's residence, is more than a mile from Moscow, Hersh has relation R to Boris Yeltsin. The abbreviation for this is the statement "Hersh R Yeltsin," or just "HRY." We might say, if we wished, "The 'ordered pair'

[H, Y] is in the relation R." This would be correct, even though in the plain English sense of the word Hersh has *no* relationship to Yeltsin—they're totally unrelated!

In mathematics, "relation" means no more than we say it means. In plain English, "relate," "relation," and "relationship," mean much more. Try not to mix up the different meanings of relation—the plain English "relation" and the mathematics "relation."

Part 1.

An "ordered pair" is by definition a set of two things, a first one and a second one. [Me, you] is an ordered pair. [You, me] is a different ordered pair.

A "binary relation on a set" is, by definition, some subset of the "ordered pairs" of elements of the set.

Example: Suppose the set in question is the set of residents of Cincinnati. Is "the grandmother of" a binary relation on the set? If Sadie and Maizie are residents of Cincinnati, and Sadie is Maizie's grandmother, then the ordered pair [Sadie, Maizie] belongs to the relation "is the grandmother of."

If R is the name of a relation, "aRb" means that the ordered pair $[a, b]$ satisfies the relation R. If a relation is such that whenever $[a, b]$ satisfies it, so does $[b, a]$, we call the relation "symmetric." ("Bidirectional" would be a more vivid name.)

1. Give examples of both symmetric and nonsymmetric relations, between people and between numbers (bidirectional and not-bidirectional).

We call a relation R "reflexive" if whenever A belongs to the set S, $[a, a]$ belongs to R. (Everything in the set S satisfies the relation R with itself.) On any set the relation of identity is reflexive; everything is identical with itself. However, identity is not the only reflexive relation. In geometry the relations of congruence and similarity are reflexive, because every figure is similar to itself and congruent to itself. The relation "lives less than a block away from" is reflexive, because everybody lives less than a block away from him/herself. In most neighborhoods it's not the identity, unless everybody lives more than a block apart.

Chapter Guidelines

2. Give other examples of reflexive and nonreflexive relations.

A relation is called "transitive" if aRb and bRc imply aRc. (The relation of "sister" is transitive. The relation "father of" is nontransitive, since the father of my father is not my father. In fact, the father of my father could be my father only if my father were his own father.)

3. Give still other examples of transitive and nontransitive relations of people and of numbers.

Part 2.

If a relation is reflexive, symmetric, *and* transitive, it's called an "equivalence relation." The subset of S consisting of all the members of S having an equivalence relation to some particular member of S is called the "equivalence class" of that member. (The relation of similarity among triangles is an equivalence relation. One equivalence class under the similarity relation is the set of all equilateral triangles, since all equilateral triangles are similar to each other.)

Show that if two equivalence classes are not identical (don't have *all* members in common), then they must be disjoint (have *no* members in common).

If we have a collection of subsets of S, and no two of these subsets overlap, and every member of S is in *some* subset, we say we have a "partition" of S.

4. Check that the relation on the integers of "having an even number as difference" is reflexive, symmetric, and transitive. By definition, it is then an equivalence relation.

5. Verify that the relation in #4 partitions the integers into two equivalence classes, the even numbers and the odd numbers.

6. CHALLENGE: Is the relation "having as a difference a number like 3, 6, 9, 30, 90, etc., which is a multiple of 3," an equivalence relation? If not, why not? If so, find out what are its equivalence classes.

Even if a relation R is not an equivalence relation, we can still consider, for any element a of S, the subset V of all elements b related to a (such that aRb).

Ch. 4: Inner Issues

Example: If S is the set of living human beings, and I, your instructor, am a, and R is the relation "is the grandchild of," then the subset V would consist of all my living grandparents. If all my grandparents are dead, V has no elements and is called "the empty set."

7. Show by several examples that if R is not reflexive, not transitive, or not symmetric, then the classes induced by R need not define a partition on S (they may not be disjoint or else may not include every element of S).

(NOTE: This is a big chunk of abstract math. It aims to show how abstraction unifies seemingly unrelated topics [kinship relations, geometric relations, numerical relations]. It could take several class periods, or could be a do-it-yourself activity, with generous class discussion after the students have several chances at it. It is meant to be fun. The last two examples point to modular arithmetic [number congruence] which return later on. The use of familiar examples from high school math begins to shed a modern light on geometry.

Emphasize the WARNING. The difficulty for some students is absorbing unfamiliar terminology and definitions. They can use help from you. Go over the definitions and examples, and let them think up one or two more in class. The whole thing, or as much of it as you like, can be done in class as a Socratic dialogue between teacher and class, or in small groups working independently during class.)

VIII. Discussion Topic: The Chinese Remainder Theorem

1. Introduction to congruence mod m (let a, b, m, be integers such that m divides $a-b$. Then we write $a \equiv b \bmod m$) as a preliminary to discussion of Shockley's version of the Chinese remainder theorem (see page 206 in the text).

2. Connections in mathematics: proof of the Chinese remainder theorem using mathematical induction. See *A First Undergraduate Course in Abstract Algebra* (2nd edition) by A. Hillman and G. Alexanderson (Belmont: Wadsworth Publishing, 1978).

Chapter Guidelines

3. Use Euclid's theorem about the infinitude of primes and the Chinese remainder theorem to prove there are a million positive consecutive integers: $x, x+1, x+2, \ldots, x+999999$, each of which is an integral multiple of the cube of a prime. See *A First Undergraduate Course in Abstract Algebra* (2nd edition) by A. Hillman and G. Alexanderson (Belmont: Wadsworth Publishing, 1978).

IX. Project Topic: The Stretched String

1. Is "straight line" a mathematical concept?

2. If yes, then when you walk a straight line are you doing math?

3. When you *think* about a straight line, is that doing math?

4. Suppose Appletown, Beantown, and Crabtown are situated on a north–south straight line. Must one be between the other two? Are you sure? How do you know? Can you prove it? Or have you ever seen it proved? What if the line were east–west? Or some other direction?

5. Now suppose the three towns are all on a *circle*, radius five miles, center at Dogtown. On that circle, must one be between the other two? Can more than one be between two others? Are you sure? How do you know? Do you think it could be proved? By what means? What if the circle were seven miles in radius, center at Hogtown?

6. A "figure eight" is a smooth, unbroken curve that has one point of self-intersection. It is traced on ice by figure skaters. If A, B, C are three distinct points on a figure eight curve, answer the same questions that you just answered for a straight line and a circle.

(NOTE: This is a "think" question. Expect a few sentences in answer to each question. There are no wrong answers except a simple yes or no. The purpose is to let the student glimpse the complexity and subtlety of seemingly simple concepts like "straight line." If you wish, you can tell the students, in appropriate language, that betweenness is a topological property, not a metric one. Here are some philosophical issues you can open, for writing assignment, class discussion, or both).

Ch. 4: Inner Issues

7. Is a straight line something we know about from observation? From a definition in a book? Is it something in our heads? What if the straight line in your head isn't "the same" as the one in my head? Could we find out? Is Euclid's straight line the same as Newton's? As Einstein's? Is the straight line of a great-grandmother who has lived all her life in a valley in the interior of New Guinea the same as Madame Curie's? If Madame Curie met her and had a common language, could she try to find out?

X. Tutorial: Complex Numbers

Complex numbers are one of the wonders of human thought. We start with an impossibility—negative numbers can't have square roots. Why not? Because a positive times a positive is positive, and a negative times a negative is *also* positive, and 0 times 0 is 0. No way to get to -1 by squaring any number! We crash through anyway, and use the letter i to stand for exactly what doesn't exist: the square root of -1. And for this courageous behavior, our reward is one of the most useful and powerful tools in mathematics!

Complex numbers are expressions such as $2 - 5i$ or $\frac{1}{2} + \frac{2}{3}i$, where i is the square root of -1. But there is no "square root of -1"!

In high school you met the quadratic equation, $ax^2 + bx + c = 0$. You learned how to solve it. In fact, the Babylonians knew how, 3,000 years ago. The two solutions or "roots" (one with a plus, the other with a minus) are

$$x = \left[-b \pm \sqrt{b^2 - 4ac}\right]/2a.$$

Stick them into the quadratic equation—they work!

What if $b^2 - 4ac$ is negative? Then the square root operation in the formula is applied to a negative number. That doesn't make sense! Negative numbers have no square roots! Yet the quadratic formula *still works*! What a puzzle—the formula is meaningless, yet it works like a charm!

Consider this simple equation:

$$x^2 + 1 = 0.$$

Chapter Guidelines

Solving this, you get

$$x = \pm(-1)^{1/2}.$$

x is plus or minus the square root of -1. But there is no square root of -1! Somebody long ago named it "imaginary" ("i" for short).

How do we operate with these imaginary numbers? The algebra of i is easy. Treat it like x, y, or any other letter. If you square it, replace it by -1. So we can perform algebraic operations on it. But still, it doesn't *mean* anything, right?

What does -1 mean *geometrically*? As a factor, a multiplier, it means "reverse direction," or "rotate through 180 degrees," or in plain English, "flip over."

Multiplying by i *twice* is the same as multiplying by -1 *once*. If multiplying by -1 means "flip over," then multiplying by i once has to be an operation that, *when applied twice*, results in flipping over.

If we are working with real numbers, with the one-dimensional number line, we cannot find such an operation. But somebody had a bright idea. Widen the context! Embed the line in a two-dimensional context, in a plane.

We ask again. What can you do *twice* that results in flipping over? In a two-dimensional context, the answer's obvious. Rotate through 90 degrees! (You could go in either direction. We mathematicians picked counterclockwise.) We apply the 90-degree rotations to 1, which we interpret as the "unit line segment"—the segment from $x = 0$ to $x = 1$, extending one unit of distance to the right, in the positive x-direction. We are interpreting i as a 90-degree rotation, so multiplying by i carries 1 off the real line up to the vertical axis. The "meaningless" symbol i is now identified with the point on the line perpendicular to the real axis at $x = 0$, one unit of distance above the x-axis. That's what a 90-degree rotation does to the unit real line segment. The second 90-degree rotation takes it to -1, which we see is indeed i^2. From one unit right of the origin, to one unit above the origin, to one unit left of the origin.

This idea pulls i out of the mysterious into the familiar. It's the basis for the tremendous power and utility of the complex

Ch. 4: Inner Issues

numbers. Three mathematicians independently found this geometric interpretation of i—Wessel, Argand, and Gauss.

What about $-i$? It must be "i flipped over"—one unit *below* the origin.

Here's another way to get $-i$. Multiply i by itself three times. You get $i^3 = (i^2)i = (-1)i = -i$. That's algebra. We can also think geometrically: operating with i three times means three counterclockwise right-angle rotations, which lands you on the downward vertical axis, which is $-i$, as before. Check!

Since i is one unit above the origin, $2i$ has to be two units up. Any real multiple of i is on the vertical axis, above or below. What about "mixed numbers," complex numbers like $2 + 3i$? The 2 says "move two units to the right." That's called "the real part." The 3 says "move three units up." That's called "the imaginary part."

We have just accomplished something awesome. Every point (a, b) in the Cartesian coordinate plane is now associated to a complex number $a + bi$, and every complex number $a + bi$ is now associated to a point (a, b). Complex numbers have capabilities that Cartesian points (ordered pairs of real numbers) don't have. They add, subtract, multiply, and divide! Thereby the points in the plane become algebraic. They can add, subtract, multiply, or divide.

What's the geometric meaning of these operations?

Addition is easy. Pick two complex numbers, say $3 + 7i$ and $2 + 4i$. Plot their points $(3, 7)$ and $(2, 4)$. The obvious, natural way to add is $(3 + 7i) + (2 + 4i) = (3 + 2) + (7 + 4)i = 5 + 11i$. Plot $3 + 7i$ and $2 + 4i$ and $5 + 11i$ in your plane. With the origin, they form a parallelogram. Another way to say it is, "the sum of two complex numbers is the fourth vertex of the parallelogram whose other three vertices are the two points and the origin."

Subtraction is no problem. It's just adding the negative of a number. You negate a number by flipping it over (multiplying by -1). So you subtract by first flipping over and then adding.

Multiplication is really exciting. We began with multiplication by i, which we interpreted as rotation counterclockwise through a right angle. To multiply a complex number by a positive real number, keep the complex number in the same direction and multiply its length by the real number. If it's a negative real number, flip over.

Chapter Guidelines

What about multiplying by a mixed number—that's neither pure real nor pure imaginary?

First look at points on the unit circle—the circle with radius one and center at the origin. Every complex number except zero has a direction and magnitude. There's exactly one complex number on the unit circle with a given direction. So every complex number is a real number times a number on the unit circle (a complex number with magnitude 1). Therefore, it's enough to be able to multiply by numbers on the unit circle—numbers of magnitude one.

A point on the circle is defined by its angle, which is customarily called θ. Its x and y coordinates are named by the two trigonometric functions, $x = \cos\theta$, $y = \sin\theta$, so the point on the circle with angle θ is $\cos\theta + i\sin\theta$. If we have a second such point with angle ϕ instead of θ, we can multiply them as a straightforward algebra exercise, and get $(\cos\theta)(\cos\phi) - (\sin\theta)(\sin\phi) + i[(\cos\theta)(\sin\phi) + (\sin\theta)(\cos\phi)]$.

We need some trigonometry, to recognize that the first two terms are just $\cos(\theta + \phi)$ and the terms in square brackets multiplied by i are $\sin(\theta + \phi)$. This formula reduces to $z(\theta) \times z(\phi) = z(\theta + \phi)$.

In words, multiplication of points on the unit circle corresponds to adding their angles—which is usually called "rotating." The complex numbers permit us to perform rotations in the plane by multiplication! This is why complex numbers are wonderful. Rotation is an important operation in mathematics and physics. To perform it by multiplying numbers is a miracle.

Next, division. A simple trick turns division into multiplication! Let's divide by $3 + 4i$. That's the same as multiplying by $\frac{1}{(3+4i)}$, which is the same as multiplying by $\frac{(3-4i)}{(3+4i)(3-4i)}$. This is the trick we warned you of—multiplying and dividing by $3 - 4i$. $(3 - 4i)$ is called the "conjugate" of $(3 + 4i)$. *Check to see that* $(3 - 4i)(3 + 4i) = 25$. In fact, any complex number except 0, when multiplied by its conjugate, gives a positive real number—the sum of the squares of the real and imaginary parts. *Prove this for yourself.* This sum of squares is called "the modulus squared." So $\frac{1}{(3+4i)}$ is equal to $\frac{(3-4i)}{25}$, which we

Ch. 4: Inner Issues

rewrite, if we like, as $\frac{3}{25} + (-\frac{4}{25})i$. Division by zero is still impossible, as it is in the reals.

There's an interesting consequence of the fact that multiplying numbers on the unit circle is done by adding angles. We can find square roots, cube roots, any roots at all! The two square roots of 1, −1 and +1, which are on the unit circle, with angles of 180 and 360 degrees. You can easily check that i, −1, −i, and 1 are fourth roots of 1. They're on the unit circle, at angles of 90, 180, 270, and 360 degrees. With these examples before us, it's easy to guess that there are *three* cube roots of 1. We already know one of them—1 itself. But there are two more, at angles of 120 and 240 degrees. They are called ω and ω^2. You can find their real and imaginary parts, either by trigonometry or by solving a certain real quadratic equation. And there are five fifth roots, equally spaced on the unit circle, six sixth roots, and so on.

The sum of the two square roots, the three cube roots, the four fourth, or the five fifth roots of any number is zero. This is easy to see in the even cases, 2, 4, etc. A proof that includes both odd and even cases is short and easy, but it requires a simple bright idea. Can you think of it?

How do you extract roots of other numbers besides 1? Let's find three cube roots of $8i$. Any cube roots of $8i$ will be the product of the cube root of 8 and a cube root of i. The cube root of 8 is 2. The imaginary unit i is on the unit circle, with angle 90 degrees. Since multiplication adds angles, one cube root has an angle of $\frac{90}{3}$ = 30 degrees. Where are the other two roots? Equally spaced around the unit circle, at 150 degrees and 270 degrees. A little trigonometry will tell you the real and imaginary parts of $2(\cos 30 + i \sin 30)$, $2(\cos 150 + i \sin 150)$, and $2(\cos 270 + i \sin 270)$.

Now back to our original question. If there's no square root of −1, how can we get away with saying there is?

The answer is, there's no square root of −1 *where we looked at first*, among the real numbers, on the real number line. But when we widened our search from the line to the plane, we found a couple of square roots.

We can say it another way. When we looked at the real line by itself, the point −1 had one coordinate: −1. But then we

49

Chapter Guidelines

started to think of the real line as an x axis embedded in the xy plane. The real line becomes the horizontal x axis of the plane. All its points have y-coordinate 0. The point we called -1 now is labeled $(-1, 0)$. This point corresponds to rotation through 180 degrees, so its square root should be rotation through 90 degrees, which has coordinates $(0, 1)$.

Algebraically, we want $(0, 1)^2 = (-1, 0)$. Write $(0, 1)$ as $0 + i$. Square it according to ordinary algebra and use the rule $i^2 = -1$.

We didn't really find a square root of the *real number* -1. We found a square root of the *complex number* -1, which is $-1 + 0i$, or $(-1, 0)$. By appropriately defining multiplication and addition of ordered pairs (complex numbers), we arranged that $(0, 1)^2$ is $(-1, 0)$. i is just an abbreviation for $(0, 1)$, and so $i^2 = -1$ after all.

XI. Project Topic: Probability

Probability is the quantitative study of randomness. It can be described as the science which asks the following question: Given a *known* collection of objects, what can be said about the characteristics of an *unknown sample* of that collection?

INSTRUCTOR: In our experience, probability is a popular topic for students. It generally is the the subject most often chosen for expository research papers. We begin with a brief outline of the basics, and conclude with a nice experiment for the classroom.

Basic terminology: An *experiment* is a situation or problem involving uncertain results. *Outcomes* are various possible results of experiments. The *sample space* is the set of all possible outcomes of an experiment. An *event* is any subset of the sample space. The *union* of two or more events is the event that at least one of them occurs. Their *intersection* is the event that all of them occur. The *complement* of an event A is the event that A doesn't occur. In a sample space S of equally likely outcomes, the *probability* of an event E is the number of outcomes in E, called $n(E)$, divided by the number of outcomes of S, $n(S)$:

$$P(E) = n(E)/n(S).$$

Ch. 4: Inner Issues

Permutations and Combinations

Permutations and combinations play a role in calculating probabilities.

Basic ideas: The number of permutations, i.e., different arrangements, of n objects taken r at a time where repetition is allowed, is n^r.

Example: How many three-letter "words" can be formed from the set of letters $\{a, b\}$? Answer: $n = 2$, $r = 3$, so $n^r = 2^3 = 8$.

The number of permutations, i.e., different arrangements, of n objects taken r at a time where repetition is *not* allowed is $n!/(n-r)!$, where $n! = (n) \cdot (n-1) \cdot (n-2) \cdots (3) \cdot (2) \cdot (1)$ and is read "n-factorial."

Example: How many ways can a two-person debate be arranged drawing speakers from an eight-member panel? Answer: $n = 8$, $r = 2$, so $n!/(n-r)! = 8!/6! = 56>$

We are not concerned with "order" or arrangement of objects when we compute the number of combinations of them. The number of combinations of n things taken r at a time is $n!/[r!(n-r)!]$.

Example: There are 10 balls in an urn, each a different color. If three balls are drawn from the urn, how many different combinations of colors are possible? Answer: $n = 10$, $r = 3$, so $n!/[r!(n-r)!] = 10!/[3!\,7!] = 120$.

Some Basic Rules of Probability

The probability of any sample space is 1.

The probability of any event is at least 0 and at most 1.

If E and F are mutually exclusive events, then the probability of their union is the sum of their separate probabilities.

Example: What is the probability of getting a black queen *or* a ten of hearts when picking one card from a deck? Answer: Let A = getting a black queen; then $P(A) = 2/52$. Let B = getting a ten of hearts; then $P(B) = 1/52$. A and B are mutually exclusive events since they cannot both happen at the same time. So the probability of their union is the sum of their probabilities: $3/52$.

Chapter Guidelines

If E and F are independent events, then the probability of their intersection is the product of their individual probabilities.

Example: A game involves tossing a fair coin and then picking a card from a deck. What is the probability of getting a head and an ace? Answer: Let A = getting a head; then $P(A) = 1/2$. Let B = getting an ace; then $P(B) = 4/52$. A and B are independent events since the outcome of one does not affect the outcome of the other. So the probability of A intersection B is the product of the probabilities: $2/52 = 1/26$.

The probability of any event E = (1 minus the probability of its complement E').

Example: What is the probability of rolling a number less than six on one throw of a die? Answer: Let E = throwing a number less than 6. Then E' = throwing a number equal to 6 (since 6 is the highest number on a die). So $P(E) = 1 - P(E')$. $P(E') = 1/6$, so $P(E) = 1 - 1/6 = 5/6$.

For *any* two events in a sample space, the probability of their union is the sum of their individual probabilities minus the probability of their intersection.

Example: What is the probability of drawing a red card or a queen in a single draw from a deck of cards? Answer: Let E = drawing a red card; then $P(E) = 26/52$. Let F = drawing a queen; then $P(F) = 4/52$. E and F are *not* mutually exclusive events, so their intersection is *not* empty. That is, there exist cards which are both queens and red—two of them. So the probability of E intersection F is $2/52$.

Thus, the probability of the union of these two events is the sum of their individual probabilities minus the probability of their intersection: $26/52 + 4/52 - 2/52 = 28/52 = 7/13$.

An Experiment

Let's conduct an experiment: Your instructor wants to divide your class into two sections. Suppose the number of students is an even number. Rather than take time and trouble with some logical method of dividing the class, the instructor decides (with the consent of the class) to simply decree that everybody whose family name starts with a letter up to or before M will be in section 1, and everybody whose family

Ch. 4: Inner Issues

name starts with N or a later letter in the alphabet will be in section 2.

There are 60 students in the class. (This number is arbitrary. You can substitute the actual number of students in your class.) When the class is divided into two sections, it turns out that there are exactly thirty students in each section.

    a. Is this a miracle?

    b. Is it just what you should expect?

    c. How would your expectations change if the problem was about an army of a million soldiers, and when split in the same manner there were exactly half a million in each sub-army?

    d. What's the extreme case in the opposite direction?

    e. How would shrinking the class change your expectation?

    f. If your instructor sets out to choose a section one classmate at a time, not knowing anyone's name, what's the chance that any one choice would have a name from A to N? If you don't know, what seems like a reasonable guess?

    g. What is the chance that your instructor would choose two in a row with names from A to N? Three in a row? Thirty in a row? Having chosen one, what is the chance that the one she picks next out of the 59 remaining would have a name from A to N. Having chosen two, with 58 remaining, what is the chance the next would have a name from A to N?

Chapter 5
Selected Topics in Mathematics

Group Theory and the Classification of Finite Simple Groups. Non-Cantorian Set Theory. Non-Euclidean Geometry. The Prime Number Theorem. Appendix A. Nonstandard Analysis. Fourier Analysis.

I. Discussion Topic: Non-Euclidean Geometry

Saccheri derived the fact that the sum of the angles of a triangle is less than 180 degrees when the parallel postulate is denied. Discuss the Saccheri quadrilateral.

II. Project Topic: Spherical Geometry

Do some investigation of spherical geometry. This is the geometry of the night sky, the geometry of celestial navigation. A spherical triangle is a region on a sphere bounded by three arcs of great circles. A meridian (circle through the north and south poles) is perpendicular (at right angles) to the equator. Consider a sphere with radius 1, and a triangle with a vertex at the North Pole and two vertices at the equator. One side of the triangles is a piece of the equator. The other two sides are pieces of meridians intersecting at the North Pole. Suppose the angle at the North Pole is n degrees. What is the angle sum of the triangle? Suppose the angle at the North Pole is very small—0.000001 degrees. Then what is the angle sum? Suppose the angle at the North Pole is almost 180 degrees, say 179.99999. Then what is the angle sum? Make a conjecture on the possible angle sums of an arbitrary spherical triangle.

The amount by which the angle sum exceeds 180 degrees is called "the spherical excess." For the triangle we are considering, it's not hard to find its area by finding what proportion of the whole sphere the triangle covers. The whole sphere has area 4π. The hemisphere has area 2π. The triangle with vertex

Ch. 5: Selected Topics in Mathematics

angle at the North Pole equal to 90 degrees is one fourth of the northern hemisphere, so its area is $\pi/2$. And so forth.

There is a beautifully simple connection between the spherical excess of any spherical triangle and its area. Find it.

Videotape Resources

Freeman, W.: "Measurement: It Started in Greece" (*For All Practical Purposes* (New York: W.H. Freeman, 1988)

Discussion Topic: Symmetries and Groups

1. Coding/decoding (cryptography) and inverses of functions

2. Transpositions in music as reflections

3. Using composition and groups to find all seven frieze patterns

4. Why do we study symmetries? How do we use algebra to combine symmetries?

5. A snail's shell is sometimes modeled by an exponential spiral (spiral of Bernoulli). Describe this spiral from the point of view of invariance. Use this example to show that a symmetry in the algebraic sense may not correspond to a symmetry in the usual visual sense.

Project Topic: Digital Sums and Casting Out Nines

Part 1.

If we add the digits of 123, we get 6. If we add the digits of 12,345, we get 15. If we then add the digits in 15, we get 6 again. In this way, from any positive whole number ("natural number") we can obtain some one-digit number, the sum of its digits or the repeated sum of digits. Let us call this number the *digital sum* of 123 or of 12,345, or of any number n, and abbreviate it "d" for short. (The term "digital sum" is not standard terminology. Since we could not find any standard term in the books, we made up our own.) Then $d(123) = d(12345) = 6$. It should be clear that if n is between 1 and 9, $d(n) = n$.

1. Choose two three-digit numbers, call them a and b. Calculate $d(a)$, $d(b)$, $d(a+b)$, $d(a) + d(b)$, and $d\bigl(d(a) + d(b)\bigr)$.

Chapter Guidelines

2. Do you notice anything interesting?

3. Would you like to make a conjecture (a guess) about what will happen for other numbers?

4. How sure are you?

5. Try another example.

6. Still another.

7. Try numbers with two and four digits instead of three.

8. How sure are you now?

9. Make an "addition" table showing $d(a+b)$ instead of the usual $a + b$ for a and b from 0 to 10 or higher.

10. Can we think of this table as defining a new, addition-like operation?

11. For this strange new addition could we try to define an "identity" (something that acts like zero under addition)?

12. Could we try to define "subtraction"?

Part 2.

13. Repeat Part 1 using the same additive definition of $d(n)$, but with multiplication and division replacing addition and subtraction in Nos. 1 and 9–12.

14. Does this new "arithmetic" follow the laws of ordinary arithmetic (associative addition, associative multiplication, commutative addition, commutative multiplication, distributive multiplication over addition)?

15. Could we calculate "square roots" and "cube roots" in this "arithmetic"?

Part 3.

16. Find an old math text that has a section on "casting out nines."

17. Read it and explain it to class.

18. How does it tie in with this problem?

(INSTRUCTOR: The students are being surreptitiously introduced to abstract algebra—they are discovering other binary operations besides the elementary ones.

Explain to the class that quotation marks are systematically used to distinguish ordinary addition and multiplication from this new, funny kind of "addition" and "multiplication."

If you're not familiar with this material, you'll find it helpful to work through the question in order to know what the students will experience. "Casting out nines" is hundreds of years old. It used to be a bookkeeper's trick to check arithmetic.)

Students demand to know how it works. The answer has two parts:

I. Taking the digital sum of a number is the same as finding its remainder on division by 9 ("modulo 9" or "mod 9," we say).

II. The sum or product of the remainders equals the remainder of the sum or product, modulo 9.

To see (I), consider a three-digit number with hundreds digit h, tens digit t, and units digit u. Write it as $(99 + 1)h + (9 + 1)t + u$. When we form the digital sum, we replace $100h = (99h + h)$ by just h, which is the same as replacing it by the remainder mod 9, since $99h$ has zero remainder.

To see (II)—why addition is preserved mod 9—add $(h + t + u)$ to $h' + t' + u'$) and compare this number with $(100h + 10t + u) + (100h' + 10t' + u')$. The two answers differ by a multiple of 9, which means they have the same digital sum. For multiplication, do the analogous thing—multiply $(h + t + u)$ times $h' + t' + u'$) and compare the result with the product of $(100h + 10t + u)$ times $(100h' + 10t' + u')$. Again the two answers differ by a multiple of 9, which means they have the same digital sum.

For division, things are more complex because 9 is factorable, which means that multiplication modulo 9 has zero divisors. It helps to look first at modular multiplication with a prime modulus like 5 or 7.

III. Discussion Topic: Nonstandard Analysis

1. In a nonstandard number system there are numbers that are not zero and yet are smaller than any positive standard number. This seems to be a contradiction. Can mathematics always create theories that "regularize" contradictory situations?

Chapter Guidelines

2. Hailed as a great advance when it was discovered, nonstandard analysis seems to have made little impact in the teaching of calculus. Discuss.

Videotape Resources

Freeman, W.: *For All Practical Purposes* (New York: W.H. Freeman, 1988): "On Shape and Size." The video opens the door to explorations of mathematical certainty, perspective, symmetry patterns, isometries in the plane, the work of Escher, non-Euclidean geometry, and fractal geometry.

"Not Knot" (The Geometry Center, University of Minnesota, 1991). 16-minute video with printed supplement.

"Outside In" (The Geometry Center, University of Minnesota, 1994). 22-minute video with printed supplement.

"Group Theory" (British Open University, No. M 101/27, 26 minutes, color, 1977). This film includes an application of symmetries to the work of post offices.

"Dihedral Kaleidoscopes," with H. S. M. Coxeter (International Film Bureau, 1966, 14 minutes); review in *Mathematics Teacher* 66 (1973).

"Mathematical Peep Show" (1961, 11 minutes); reviewed in *Mathematics Teacher*, November 1971, p. 625.

"Symmetries of the Cube" (International Film Bureau, 1971, 14 minutes, color); reviewed by Everett Van Akin in *Mathematics Teacher*, No. 8, 1972, p. 733.

Freeman, W.: *For All Practical Purposes* (New York: W.H. Freeman, 1988): "Scale and Form: How Big is Too Big?" 30 minutes, includes a discussion of symmetries, frieze patterns, Escher, and more.

Freeman, W.: *For All Practical Purposes* (New York: W.H. Freeman, 1988): "Computer Science: Rules of the Game": algorithms; "Counting by Twos": numerical representation; "Creating a Code": encoding information; "Moving Picture Show": computer graphics.

Chapter 6
Teaching and Learning

Confessions of a Prep Teacher. The Classic Classroom Crisis of Understanding and Pedagogy. Pólya's Craft of Discovery. The Creation of New Mathematics. An Application of the Lakatos Heuristic. Comparative Aesthetics. Nonanalytic Aspects of Mathematics.

I. Discussion Topic: Problem Solving

1. How is the process of generating ideas different from the process of evaluating them?

2. Some advice for solving mathematical problems based on the philosophy of George Pólya:

A. Make Reasonable Guesses

- Try to make guesses that help narrow the scope of the problem, like order-of-magnitude guesses.
- Be sure you can give reasonable evidence of the bases for your guesses.
- Maintain a healthy skepticism toward your guesses.
- Investigate the consequences of your guesses; make new guesses in light of new insights.
- Keep a record of all determinations you make as you progress.

B. Solve Simpler Cases

- Test extreme cases (extremely small or large).
- Consider *all* parameters that can reasonably be varied.

C. Look for Patterns in the Data

- Use any perceived pattern to make a prediction.
- Check your prediction. If it's incorrect, look for another pattern. If it is correct, write a description of the pattern on which it is based.
- If at least two predictions check out, try to *prove* that the pattern on which they were based is general.

Chapter Guidelines

II. Project Topic: Magic Sevens and Periodic Decimals

Part 1.

Let n denote the six-digit number 142,857.
   a. Calculate $2n$ and $3n$.
   b. Do you notice anything funny?
   c. Are you willing to guess in advance what happens next?
   d. Write down your guess, then check it by calculating $4n$.
   e. Guess again, and check again with $5n$.
   f. Guess and check one after the other $6n$, $7n$, $8n$.
   g. Are you a good math guesser? Give yourself a letter grade.

Part 2.

   h. Carefully and accurately, *by hand* divide 7 into 1 to eight decimal places.
   i. Does this part have anything to do with Part 1?
   j. What?
   k. How?
   l. Why?
   m. Guess what will come in the 10th, 16th, and 22nd decimal place. *Write down your guess.*
   n. Check.
   o. Give yourself a letter grade for guessing.

Part 3.

   p. Calculate 1/9, 1/11, and 1/13 to enough decimal places so that you can guess correctly what would appear in the 30th place (if you felt like carrying the calculation that far).
   q. Check.
   r. Experiment with the six-digit repetend you get from 1/13.
   s. The term "repetend" has not been defined. It may not be in your dictionary. Guess the meaning from the context and the root word, "repeat," or ask your instructor.
   t. Try multiplying it by numbers 2, 3, 4, etc.
   u. Figure out what's going on.
   v. Guess in advance the result of multiplying it by 12 and 13.

Ch. 6: Teaching and Learning

w. Explain why the repetend can have six digits rather than five or seven.

INSTRUCTOR: We always show the class why (A) every rational number is equal to an (ultimately) repeating decimal and (B) every repeating decimal is equal to a rational number. (For example, $1/12 = 0.0833333$ is not quite a repeating decimal, but it is *ultimately* repeating.) For (A), as the division is carried out, they always obtain a remainder less than the divisor, so if the divisor is n, in at most n steps the remainder must repeat, and from then on the whole calculation must repeat. For (B), we give a little lesson on geometric sequences and series, starting with Zeno's paradox of Achilles and the tortoise, and the series $1/2 + 1/4 + 1/8 + \cdots$.

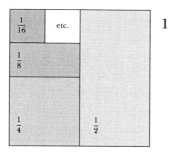

Geometric representation of Zeno's paradox.

(Don't define "limit" explicitly. Just note that if s is the sum and r is the ratio (in this example $1/2$), then $(1/2)$ times $s = s - (1/2)$. Let them solve this equation for s. Be sure to consider $r = 1$, $r = -1$, $r = 10$, to show that $|r|$ less than 1 is a necessary condition, and more important, to show that *a plausible conjecture, even if it seems to check, often is false without some unforeseen, necessary condition*. This deserves a whole lesson of its own. The application to repeating decimals is an unexpected, good review.)

(This problem can be a first look at number theory, or a second look after the problem on congruence mod 9. This one is closer to familiar ideas, but more demanding computationally. In general, we welcome calculators or computers.

Chapter Guidelines

the student to notice the calculation itself, not just the
result.)

(You could go on to 1/17, 1/19, 1/21, and so on. Or try to
explain why the permutations are cyclic and all of the same
length, and why that length is what it is. You could transpose
the whole thing to a base other than 10 and compare the
results of the two investigations.)

III. Project Topic: EXPLORE:*
Design a Winning Strategy for a Numerical Game

Rules of the game:

1. Four players choose two teams. Each person on each team will be recording. At the top of your papers, write the name of each person in the group.

2. As a group, choose a whole number between 1 and 100, at random. Call it N.

3. Team 1 chooses a positive integer different from N that is a divisor of N (i.e., that divides evenly into N so that the remainder is zero), and subtracts that divisor from N. The result of the subtraction is given to Team 2.

4. Team 2 works with the results of the subtraction just as Team 1 worked with N: choosing a positive divisor of that number different from the number itself, and subtracting the chosen divisor form it. The result of the subtraction is given to Team 1.

5. Continue the play, with teams taking turns choosing divisors and subtracting until the result reaches the number 1. The team that produces the result 1 wins.

The goal of this group project is to design a winning strategy for EXPLORE.

Directions for the Classroom Groups

1. Play three games of EXPLORE. Keep track of your games, i.e., keep records so you will have a collection of data to use

---

* Adapted from *Foundations of Higher Mathematics* by Fendel/Resek, 1990.

Ch. 6: Teaching and Learning

in designing your winning strategies. You'll want to look for patterns among your winning examples.

2. As a group, formulate at least two questions about a winning strategy for the game and write these questions down. Then discuss the questions in your group and try to conjecture about possible answers.

3. Test your conjectures by playing at least three more games.

4. Discuss the results with your group and write down your conclusions.

After You Have Played EXPLORE**:**

Answer each of the following questions completely:

a. Is there always a winner in EXPLORE? Write a statement explaining your answer. Identify any basic facts about whole numbers that you need.

b. What are the possibilities for the last move of EXPLORE? Why are there no other possibilities? Explain carefully, identifying any basic facts about the whole numbers that you need.

c. Give a complete winning strategy for EXPLORE for the appropriate team, using *each* of the numbers from 2 through 10 as the starting number.

d. Try to determine how many turns a game of EXPLORE might take. Write down answers to the following to help you answer these questions:

1. In terms of the starting number, what is the largest number of turns the game can take?

2. In terms of the starting number, what is the smallest number of turns the game can take?

3. In terms of a specified number of turns, what is the smallest number you can start with?

4. In terms of a specified number of turns, what is the largest number you can start with?

Videotape Resource

Pólya, George: "Let Us Teach Guessing" (Washington D.C.: The Mathematical Association of America, 1965).

Chapter 7
From Certainty to Fallibility

Platonism, Formalism, Constructivism. The Philosophical Plight of the Working Mathematician. The Euclid Myth. Foundations, Found and Lost. The Formalist Philosophy of Mathematics. Lakatos and the Philosophy of Dubitability

I. Discussion Topic: Platonism, Formalism, Constructivism

1. Platonism views mathematics as a study about eternally given objects which impose their nature upon the results of mathematics. Mathematics thus deals with abstract or ideal "objects" which have *been* independent of our thought about them. Discuss π, the Pythagorean theorem, the Fibonacci sequence, etc., from the viewpoint of the Platonist. Talk about Gödel and his connection to Platonism.

2. Constructivists admit only the existence of mathematical objects and theories that can be constructed. Constructivists emphasize the *algorithmic* aspects of mathematics. Discuss the avoidance of the infinite in the mathematics of the ancients. Make connections between the constructivist's views of mathematics and the needs of computer science.

3. Formalists reject the notion that mathematics relies on experience. There are no mathematical objects—just formulas. Discuss the move to formalism that occurred during the early twentieth century and what concerns motivated mathematicians in this direction.

II. Discussion Topic: The Euclid Myth

1. The nineteenth-century discovery of the independence of the parallel postulate destroyed the view of Euclidean geometry as the exemplar of truth, freed mathematics from a

total reliance on spatial intuition, and paved the way for a flourishing of many new fields of geometry.

2. The destruction of the Euclid myth gives evidence of mathematics as process and not a mere collection of facts or "eternal truth."

3. New branches of mathematics are often the consequence of investigating familiar facts. Thus investigations into the parallel postulate of Euclid led to the birth of new branches of mathematics like non-Euclidean geometries. A similar situation occurred in the birth of nontraditional algebras in connection with investigations into the commutativity of multiplication.

4. What is possible in one context in mathematics is often impossible in another. Consider: lines that sometimes intersect (Euclidean geometry); lines that always intersect (Riemannian geometry); and lines that never intersect (Lobachevskian geometry). Study operations that maintain certain properties on one algebraic system (commutative multiplication of the integers) but lose them in another (noncommutative multiplication of quaternions).

5. Exploring the idea that "the familiar can be a source of new knowledge" one can illustrate the multiculturalism of mathematics. Discuss frieze patterns and the connections between the artifacts of different cultures and the Pythagorean theorem. See for example "A Widespread Decorative Motif and the Pythagorean Theorem" by P. Gerdes in *For the Learning of Mathematics*, vol. 8, 1988.

6. Gödel's incompleteness theorem is another example of a result that stunned the mathematical world. Explore the origins of this theorem in the ancient Greek (liar) paradox of Epimenides ("This statement is false") showing how Gödel simply translated it into mathematical terms, using mathematical reasoning to explore mathematical reasoning itself. See for example *Gödel's Proof* by Ernest Nagel and James R. Newman (New York University Press, 1960).

Videotape Resources

Freeman, W.: "Measurement: It Started in Greece" (*For All Practical Pur-

Chapter Guidelines

poses (New York: W.H. Freeman, 1988) has nice illustrations of models of Riemannian and Lobachevskian geometries.

III. Discussion Topic: What Is Mathematics and What Do Mathematicians Do?

1. *What do mathematicians do?* Interview a practicing mathematician (academic, industrial, statistical, computer, etc.). See how she/he answers this question about her/his work.

IV. Discussion Topic: The Axiomatic Method

1. The building blocks of deductive mathematics are the axioms. What is the power of the axioms in defining a mathematical theory?

2. Axioms can be used to illustrate the connections between different branches in mathematics. They provide a structure for understanding how different parts of mathematics are related.

3. Axioms provide an opportunity for the growth of mathematics. The axiomatization of geometry in ancient Greece transformed mathematics from an experimental science into an intellectual one. Several major crises, revolutions, directional changes, and developments in the history of mathematics were tied to axioms.

4. Certain axioms have played an important role in the history of mathematics. It is fascinating to try to determine which axioms are crucial to which mathematical theories. It is also interesting to determine what consequences result from denying specific axioms.

5. In 1904, Professor Maxime Bôcher claimed: "Until a system of axioms is established, mathematics cannot begin its work." But 80 years later, Professor Morris Kline said, "When a mathematical subject is ready for axiomatization, it is ready for burial and the axioms are its obituary." How can we evaluate each of these statements within their historical context and come to some opinion of their accuracy in today's mathematical world?

Chapter 8
Mathematical Reality

The Riemann Hypothesis. π. Mathematical Models, Computers, and Platonism. Classification of Finite Simple Groups. Four-Dimensional Intuition. True Facts About Imaginary Objects. Why Should I Believe a Computer?

I. Discussion Topic: Four-Dimensional Intuition

Goal: Help students understand the "possibilities" in mathematics, and the ability of mathematicians to reason intuitively in contexts beyond physical experience.

1. Explore the many dimensions of mathematics. In E. Abbott's *Flatland: A Romance of Many Dimensions* (Princeton: Princeton University Press, 1991), one can experience the restrictions of living in a two-dimensional world.

2. Artists who wanted to depict our three-dimensional world on canvas gave impetus to the birth of projective geometry.

3. Mathematical facts and formulas are often born in a certain dimension and are extended to other dimensions. Consider the Pythagorean theorem, which is stated in a two-dimensional environment. How do we extend it to three dimensions? See "A New Look, Pythagoras" by C. Thornton in *The Mathematics Teacher*, vol. 74, no. 2, February 1981.)

4. Are there mathematical formulas that hold in two dimensions, but not in three?

5. We can obtain "evidence" for mathematical results in two- and three-dimensional environments. Is this possible in four dimensions? What process do we use to make mathematical statements about a four-dimensional environment?

Chapter Guidelines

II. Project Topic: A Round Trip to the Fourth Dimension

Part 1.

You know three-dimensional cubes and two-dimensional cubes (usually called "squares").

1. Does there exist a four-dimensional cube? If there is such a thing, how many parts does it have?

Comment: At first this problem doesn't make sense, for several reasons. You may respond, "How can you ask me about the fourth dimension? I don't have a clue what that means." But in a few minutes you'll be answering questions about the four-dimensional cube! (For brevity, we say "hypercube" instead of "four-dimensional cube.")

This is a good candidate for Professor George Pólya's famous principle of problem-solving: *If you can't solve the problem you have, think of a related problem you might be able to solve.*

2. What is a related problem that might be easier to solve?

3. How many parts has an ordinary cube (a 3-cube)?

4. What are "parts"?

For purposes of this problem, the parts of a cube are its interior (three-dimensional), its faces (two-dimensional), its edges (one-dimensional), and its vertices or corners (zero-dimensional).

5. Count the various kinds of parts of a cube, write down the numbers in order in a row, and add them up.

6. It's not clear how to go up to four dimensions. But it would be easy to go _____ (fill in the blank).

7. How many parts has a 2-cube (square)? Count them up, write them down in order in a second row above the first row, and add them up.

8. What should we understand by a 1-square? How many parts has it, of which kinds? Write them down in a row above the second row.

9. Same instructions for the zero-square. Now you have a table, composed of four rows neatly going from zero to three.

10. Study the numbers in front of you till you see what should go in the fifth column for the hypercube. (There is

Ch. 8: Teaching and Learning

a simple relation between the numbers in any row and two numbers in the previous row.)

11. How many parts has a hypercube?

Part 2.

12. Can you extend the table to more than five rows and more than five columns?

13. What would appear in the fourth row, k-th column? Fifth row, k-th column? Sixth row, k-th column?

14. Guess what would be the sum of the numbers in the n-th row.

15. Prove that the sum of the parts of an n-cube is 3^n in two different ways: by induction using the recursion formula, or by the binomial theorem, summing the number of parts of the n-cube of dimension k.

A triangle is sometimes called a 2-simplex, because it's two-dimensional and is the simplest figure in two-space. Similarly, a tetrahedron (triangular pyramid) is called a 3-simplex.

16. How many parts has a 4-simplex?

Part 3.

A square can be constructed by connecting two parallel line segments. A 3-cube can be constructed by connecting two parallel squares. Construct a hypercube from two 3-cubes, and then from this recursive construction give the geometric explanation of the recursion formula you discovered empirically in Part 2.

Part 4.

We can use coordinates to study cubes combinatorially. A 1-cube (unit line segment) can be taken as the segment of the x-axis between $x = 0$ and $x = 1$. A unit square can be given vertices in the x–y plane as $(0,0)$, $(0,1)$, $(1,0)$, and $(1,1)$. A unit cube in the x–y–z space has vertices at points whose three coordinates are all either 0 or 1. Verify that there are eight such vertices. So a unit hypercube is defined as the region in four-dimensional (x–y–z–w) space where the vertices have *four* coordinates that are either 0 or 1.

Chapter Guidelines

An *edge* in any number of dimensions is defined by two vertices, all of whose coordinates are the same except one. (For example, in four-dimensional space, one edge of the hypercube connects the vertex at $(0,0,0,0)$ and the vertex at $(0,0,0,1)$.) A two-face is defined by a pair of *edges*, all of whose coordinates are the same except for two. For example, in three dimensions the face of the unit cube in the x–y plane ($z = 0$) can be defined by two parallel edges, the first on the x-axis, the second one unit up in the y-direction ($y = 1, z = 0$). A three-face (in four-space) is defined by two parallel two-faces, just like the three-cube in three-space.

17. Using these definitions, recalculate the number of parts for the 2-, 3-, and 4-cube.

Part 5.

18. Now do you believe there is such a thing as a four-dimensional cube?

19. If not, then how come we are able to find out so many facts about it?

20. If yes, then where and how does it exist?

(Note to the Instructor: The philosophical questions at the beginning and end of the lesson are the main reasons for doing it. Students should begin to get an idea of how far and how reliably mathematics can stretch their imaginations. Of course, you may want to assign only part of this problem.

The number of k-parts of the n-cube is 2^n times $\binom{n}{k}$, or $\frac{2^n (n!)}{k!(n-k)!}$. Summing over k, the binomial theorem gives 3^n.

There may be questions about what this has to do with *time* as the fourth dimension in relativity theory. The shortest and simplest answer is, "Nothing."

In one class a student commented, "This is really *pure* mathematics, isn't it?" I took this to be a challenge: "What's the use of all this anyhow?" Fortunately, a partial answer is easy to give and very worthwhile in itself. The preeminently practical subject of linear programming deals with n-dimensional linear geometry. True, we aren't mainly concerned with counting parts. But it is n-dimensional.

Ch. 8: Teaching and Learning

After all, a mathematical square or cube doesn't exist physically any more than a hypercube. Yet all three are interesting and useful.)

Videotape Resources

Banchoff, Thomas and Strauss, Charles: "The Hypercube" (Chicago: International Film Bureau, Inc., 1978).

Apostol, Tom: "The Theorem of Pythagoras" (MATHEMATICS!: Pasadena: California Institute of Technology, 1988).

Freeman, W.: "Measurement: It Started in Greece" (*For All Practical Purposes*, New York: W.H. Freeman, 1988)

II. Discussion Topic: Why Should I Believe a Computer?

1. What role does the computer play in mathematical proof?

2. Can the computer be considered a tool of mathematical proof? Discuss the role that different tools have played in establishing mathematical results. For example, certain problems (the trisection of an angle) are unsolvable using only Euclidean tools (straightedge and a compass), but solvable when Euclidean tools and the conchoid are utilized. The four color theorem is unsolvable (as yet) without the computer, yet accessible with the computer.

Part III
Sample Syllabus

Sample Syllabus

What follows is a sample syllabus for a three-unit general education course, "Mathematical Ideas." Prerequisites for the course are measurement geometry and intermediate algebra. Students are generally those in non–mathematics related majors and number 30 per class. Classes are scheduled for 15 weeks, 3 hours weekly.

First Day Handout

> Mathematics 131: Mathematical Ideas
> Spring, 1995   Tuesday/Thursday 11:00–12:15
> Text: *The Mathematical Experience*

Course Objectives

The goal of this course is to give you a sense of what mathematics is and what mathematicians do. Course topics include history and philosophy of mathematics as well as mathematics.

Basis for Grading

Homework Assignments:	15%
Midterm:	20%
Final:	25%
Paper:	25%
Class Participation:	15%

Homework Assignments

Homework assignments will consist of problem sets, readings, and essays. No late homework will be accepted. Some assignments will require use of the Mathematics Resource Library which is located in the Learning Resource Center, Room 610, Engineering Field. This library was especially designed for this course. The library is open M–F 8 a.m. to 5 p.m.

Paper

You are asked to write a 5-page expository research paper as a class project. We will have class discussions about possible topics. The following are deadlines for this project:

Sample Syllabus

February 23:	Topic Selection
March 7:	Bibliography due
March 16:	Outline due
March 30:	First Draft due
April 27:	Final Paper due

Class Participation

This is a student-centered, rather than instructor-centered, class. To that end, we will form small groups to work on assigned problems or discuss assigned reading. Grade will be based on your preparedness and participation in these activities, and performance on group projects.

Syllabus

INSTRUCTOR: One of the benefits of using *The Mathematical Experience* is that material from chapters can be introduced in different orders. This syllabus demonstrates one possible sequence.

Rationale

We begin this class with a discussion of what mathematics is and what mathematicians do. The first focus is "The Mathematician as Prover." This enables the instructor to introduce the student to the distinctions between evidence and proof, to understand the role of conjecture and counterexample in creating mathematics, and to experience different ways of proving. A discussion of Abstraction, Generalization, and Formalization leads very nicely into the second focus, "The Mathematician as Pattern Finder." For example, when students try to discover a generating formula for Pythagorean triples, they collect evidence, abstract information and generalize to make conjectures about their generating formula. The final step is proof that their conjectures are valid.

Week I

Day One

IN CLASS:

1. Getting acquainted, distribute class syllabus.

Mathematics 131: Mathematical Ideas

2. Classroom Discussion: The alien has landed. Chapter 1, Question 1, page 11 in this *Companion Guide*.

3. Have the students write a brief mathematical autobiography during class.

HOMEWORK:

Read the text to page 4. Due next class meeting.

Day Two

IN CLASS:

1. Discuss research paper, going over deadlines listed on handout. Visit library.

2. Group Work Activity: In your own words, describe the task or goal the authors envision for the text. What is Gian-Carlo Rota's observation about oversimplification as it relates to mathematics? What, in your opinion, would be an oversimplified view of mathematics? Instructor: see page 89 of this guide.

HOMEWORK:

Read Chapter 1. Due next class meeting.

Essay Assignment: Chapter 1, Number 1 (page 31 in the text). Due in one week. Typed.

Theme, Weeks II to IV: The Mathematician as Prover

Week II

Day One

IN CLASS:

1. Group Work Activity: What is Ulam's dilemma? If you were Ulam, and a reporter from Newsweek was interviewing you regarding this dilemma, how would you explain it? Is there anything that can be done about Ulam's dilemma? Will the Information Superhighway help solve this dilemma? Explain. Instructor: see page 90 of this guide.

Sample Syllabus

2. Class discussion: Proof, verification, conjecture, evidence, intuition. What do these words mean and how do they relate to one another? Goldbach Conjecture. Fermat's Last Theorem.

HOMEWORK:

Read pages 36–48. Due next class meeting.

Go to the Mathematics Media Lab and view the videotape "The Theorem of Pythagoras."

Day Two

IN CLASS:

1. Classroom Discussion: Different types of proof. Proof by mathematical induction; proof by infinite descent (Fermat); counterexample (see this *Companion Guide*, pages 12–13 in the guidelines for Chapter 2).

2. Group Work Activity: In *The Ideal Mathematician* what particular "difficulties of communication emerge vividly" from the exchange between the ideal mathematician and the *public relations officer*? Can you find any specific evidence of contradiction between what the ideal mathematician believes and what he can explain to the *student*? See page 91 of this guide.

HOMEWORK:

Read pages 59–69. Due next class meeting.

Week III

Day One

IN CLASS:

1. Classroom Discussion: Proof by contradiction. Dissection proofs of Pythagorean theorem. (Students had observed these in the videotape assigned last week.)

2. Group Work Activity: Create "pieces" for a dissection proof of the Pythagorean theorem and give each group the pieces, asking them to recreate the proof of the Pythagorean theorem.

HOMEWORK:

Read pages 87–97. Due next class meeting

Day Two

IN CLASS:

1. Classroom Discussion: Direct proof (algebraic) of the Pythagorean theorem (students had observed this in the videotape assigned in Week II).
2. Group Work Activity: Dissection experiment in regards to square/rectangle (see Chapter Guidelines Topic II, Question 2, page 13).

HOMEWORK:

Read pages 138–167.

Week IV

Day One

IN CLASS:

1. Classroom Discussion: Abstraction, Generalization, Formalization (Chapter 4). Euclid's proof of the Pythagorean theorem compared to other proofs we have discussed.
2. Group Work Activity: Generalize this statement in two different ways: If the sides of a rectangle have length a and b, its area is ab. Consider the following two statements and decide whether B is a generalization of A. Explain your reasoning.

   Let A be the statement: The medians of any triangle intersect in a single point.

   Let B be the statement: The angle bisectors of any triangle intersect in a single point. (See page 93 of this guide.)

HOMEWORK:

Read pages 188–195.

Essay assignment: Chapter 4, page 218 in the text, Question 3: Mathematics is the subject in which there are proofs. Explain to your younger sister different types of proofs that mathematicians use, and try to give her a sense of which,

Sample Syllabus

if any, have been most convincing from your perspective. Use specific examples. Due in one week.

Day Two
IN CLASS:

1. Classroom Discussion: Prove the bisectors of the angles of a triangle are concurrent. This follows from the group activity of day one.

HOMEWORK:

Extra Credit Assignment: Prove the medians of a triangle are concurrent.

Theme, Weeks V to IX: The Mathematician as Pattern Finder

Week V

Day One
IN CLASS:

1. Classroom Discussion: Give examples of order out of order, chaos out of order, order out of chaos, chaos out of chaos. Instructor: see page 188 in the text.
2. Discussion of Research Papers

HOMEWORK:

Work on the Research Paper. Bibliography due next class meeting.

Day Two
IN CLASS:

1. Classroom Discussion: Try to determine a generating formula for primitive Pythagorean triples. (See "The Pythagorean Problem" in *Invitation to Number Theory* by Oystein Ore; see page 133 in the text.)

HOMEWORK:

Go to the media lab and look at the videotape *For All Practical Purposes*: "On Shape and Size" (30-minute videotape). The Introduction includes a discussion of Fibonacci numbers and the golden ratio.

Mathematics 131: Mathematical Ideas

Week VI

Day One

IN CLASS:

1. Classroom Discussion: Review the students' conjectures for generating formula for Pythagorean triples.
2. Group Work Activity: Use the conjectures to determine generating formulas. See page 95 of this Guide.

HOMEWORK:

Work on the Research Paper. Check out books, articles suggested by instructor.

Day Two

IN CLASS:

1. Classroom Discussion: Chapter 3, Topic III, Question 2, page 33 in this Guide. Leonardo of Pisa (c. 1202) developed the Fibonacci sequence in trying to answer the following question: If someone places a pair of rabbits in a certain place enclosed on all sides by a wall, how many pairs of rabbits will be born there in the course of one year, it being assumed that every month a pair of rabbits produces another pair, and that rabbits begin to bear young two months after their own birth? Describe how the Fibonacci sequence answers this question.
2. Classroom Activity: Make a conjecture about the sum of the the first n Fibonacci numbers.

HOMEWORK:

Make a conjecture about the sum of Fibonacci numbers with even subscripts. Make another conjecture about the sum of Fibonacci numbers with odd subscripts. Give evidence to support your conjectures.

Week VII

Day One

IN CLASS:

1. Classroom Discussion: Prove the students' conjectures about the Fibonacci numbers using mathematical induction.

Sample Syllabus

HOMEWORK:

Read pages 97–116.

Long-term Essay Assignment: Chapter 3, question 19, page 130 in the text. Due at close of discussion of Fibonacci numbers.

Day Two

IN CLASS:

1. Classroom Discussion: Revisit the dissection proof with regards to the square/rectangle from Week 3. Show how Fibonacci numbers can be used to describe the one unit discrepancy when you dissect an 8-inch square and try to form it into a 5×13-inch rectangle.
2. Group Work Activity: Golden Rectangle (Problem 4 of Chapter 3, page 131 in the text.

HOMEWORK:

Read pages 184–187.

Week VIII

Day One

IN CLASS:

1. Classroom Discussion: The Golden Ratio and Fibonacci Numbers.
2. Group Activity: The Golden Ratio and Fibonacci numbers. See page 97 of this Guide.

HOMEWORK:

Work on Research Paper.

Day Two

IN CLASS:

1. Classroom Discussion: Fibonacci Numbers and Pythagorean triples.

HOMEWORK:

Work on Research Paper.

Mathematics 131: Mathematical Ideas

Week IX

Day One

IN CLASS:

1. Classroom Discussion: Pascal's triangle.
2. Group Activity: Pascal's triangle. See page 96 of this Guide.

HOMEWORK:

Work on Research Paper.

First Draft due next class meeting.

Day Two

IN CLASS:

1. Classroom Discussion: Fibonacci numbers and Pascal's triangle (Problem 9 of Chapter 4, page 221 in the text). Essay Assignment: Fibonacci numbers (question 19 of Chapter 3, page 130 in text.) Due in two weeks.

Week X

Day One

Review for Midterm Exam.

Day Two

Midterm Exam.

Theme, Weeks XI to XV: Geometry, An Ever Fruitful Product of the Mathematician's Curiosity and Imagination

Week XI

Day One

First draft of research paper due; return midterm examination.

IN CLASS:

Classroom Discussion: Platonism, formalism, constructivism; Chapter 7.

Sample Syllabus

HOMEWORK:

Read pages 356–368. Look at videotape: "Measurement: It Started in Greece" (*For All Practical Purposes*). This has nice illustrations of models of Riemannian and Lobachevskian geometries.

Essay Assignment: The destruction of the Euclid myth gives evidence of mathematics as process and not a mere collection of facts or "eternal truths." Write an essay for the *Atlantic Monthly* alerting the world to the view of mathematics as an evolving discipline and give specific examples to support this view. Due in one week.

Day Two

IN CLASS:

1. Classroom Discussion: Geometry and Intuition. Freewrite: What is the Euclid Myth? Chapter 7.
2. Group Work Activity: Platonism, formalism, constructivism. See page 98 of this Guide.

HOMEWORK:

Read pages 241–247; 433–441.

Week XII

Day One

IN CLASS:

1. Classroom Discussion of non-Euclidean geometries. Demonstrate in hyperbolic geometry that there exists a triangle with angle sum less than 180 degrees; Chapter 5.

HOMEWORK:

Read pages 85–87, "Mathematical Models."

Day Two

IN CLASS:

Classroom Discussion: Finite Geometries and Models. For reference, see for example *A Course in Modern Geometries* by Judith N. Cederberg (New York: Springer-Verlag, 1989).

Mathematics 131: Mathematical Ideas

Axioms for a four-point geometry:

Primitives: point, line, on.

Axiom 1: There exist exactly four points.

Axiom 2: Two distinct points are on exactly one line.

Axiom 3: Each line is on exactly two points.

HOMEWORK:

Read pages 377–397. Create two models for a three-point geometry based on the following undefined terms and axioms:

Primitives: point, line, on.

Axiom 1: there exist exactly three points.

Axiom 2: two distinct points are on exactly one line.

Axiom 3: not all points are on the same line.

Axiom 4: two distinct lines are on at least one common point.

How many lines exist in this geometry? Discover and prove two theorems for this geometry.

Week XIII

Day One

IN CLASS:

Show and discuss videotape: *The Hypercube* (Banchoff and Strauss).

Group Work Activity on the Hypercube (see page 99 of this Guide).

HOMEWORK:

Read pages 442–453.

Day Two

IN CLASS:

Begin Project: "A Round Trip to the Fourth Dimension" (Instructor: see Chapter Guidelines for Chapter 8 in this guide).

Sample Syllabus

HOMEWORK

As required by Project.

Week XIV

Day One

Research Paper due; complete Project.

Day Two

Sample Final distributed.

Week XV

Review for Final Examination.

Part IV
Sample Group Activities

Sample Group Activities

Introduction and Ch. 1: The Mathematical Experience

Names:

1. In your own words, describe the task or goal the authors envision for *The Mathematical Experience*.

2. (a) What is Gian-Carlo Rota's point about oversimplification as it relates to mathematics?

2. (b) What, in your opinion, would be an oversimplified view of mathematics?

Companion Guide to The Mathematical Experience

Ch. 1: The Mathematical Landscape

Names:

1. What is Ulam's dilemma? If you were Ulam, and a reporter from *Newsweek* was interviewing you regarding this dilemma, how would you explain it?

2. Is there anything that can be done about Ulam's dilemma?

3. Will the Information Superhighway help solve this dilemma? Explain.

Sample Group Activities

Ch. 2: Varieties of Mathematical Experience

Names:

1. In *The Ideal Mathematician*, what particular "difficulties of communication emerge vividly" from the exchange between the ideal mathematician and the public relations officer?

2. Can you find any specific evidence of contradiction between what the ideal mathematician believes and what he can explain to the student?

3. Describe the tone of this essay.

Companion Guide to The Mathematical Experience

Ch. 3: Outer Issues: Utility

Names:

1. When we talk about the utility of mathematics to mathematics, what do we mean? Give two examples that we have discussed in class, demonstrating the utility of mathematics to mathematics.

2. What do the authors mean by common utility of mathematics? Give an example demonstrating this phenomenon.

3. What is the difference between Hardyism and Mathematical Maoism?

Sample Group Activities

Ch. 4: Inner Issues

Names:

1. *Generalize* this statement in two different ways:

If the sides of a rectangle have length a and b, its area is ab.

2. Consider the following two statements and decide whether B is a *generalization* of A. Explain your reasoning.
Hint: Draw a picture.

Let A be the statement: *The medians of any triangle intersect in a single point.* (Note: a median of a triangle is a line joining a vertex to the midpoint of the side opposite the vertex.)

Let B be the statement: *The angle bisectors of any triangle intersect in a single point.* (Note: An angle bisector of a triangle is the line that divides the angle into two equal angles.)

Companion Guide to The Mathematical Experience

Ch. 3: The Pythagorean Theorem

Names:

1. What are *Pythagorean triples?*

2. Describe to a student who has not attended this class the concept of a *dissection proof.*

3. The videotape "The Theorem of Pythagoras" uses a *shearing* process in the dissection proofs. What is the effect of *shearing* a triangle on the calculation of its area ?

4. The Pythagorean theorem states a fact about a right triangle in a plane. Where does the Pythagorean theorem *not* hold, and why does it *not* work in that environment?

5. Describe two real-life situations that model the Pythagorean theorem.

6. You are a professor trying to decide whether or not to show the videotape "The Theorem of Pythagoras" to your class. Make your decision, and then support it. That is, if you decided to show it, describe at least one mathematical fact that you hope your students will learn from it. If you decided not to show it, criticize its treatment of at least one specific topic.

Sample Group Activities

Ch. 3: Pythagorean Triples

Names:

We have made the following conjectures for triples (a, b, c) where $a < b < c$:

Dawn: Every primitive Pythagorean triple has only one even number.
Amy: When a is odd, $c - b = 1$.
Rob: All primitive contain at least one prime number.
Denise: b is divisible by 4.
Rick: Every odd number a is in a Pythagorean triple.
Steve: When a is odd, we can generate a subsequent triple:

$$(a + 2, 2a + c + 1, 2a + c + 2)$$

We were attempting to prove that Steve's conjecture was true. We discovered it would be, under the given conditions, if it is always the case that $2c = a2 + 1$. Consider the triple $(5, 12, 3)$. Does $2(13) = 5^2 + 1$? The list of triples we discovered:

$(3, 4, 5)$ $(8, 15, 17)$ $(5, 12, 13)$ $(7, 24, 25)$
$(9, 40, 41)$ $(11, 60, 61)$

1. Can you simplify *Steve's* conjecture? Hint: In addition to what we've proved, use the conjectures made by *Rick* and *Amy*.

2. Try to generate a formula for triples where a is even. This may even turn out to be a generating formula for all triples. Hint: Since we conjecture (thanks to *Rob*) that every triple contains an even number, try rearranging your triples as follows: (a, b, c) where a is even and $b < c$. Try to write $a, b,$ and c in terms of two other integers m and n.

Companion Guide to The Mathematical Experience

Ch. 4: Pascal's Triangle

Names:

Perform the following tasks:

1. We have solved several problems using Pascal's triangle. Try to describe what "type" of problem lends itself to solution by Pascal's triangle.

2. Create your own problem that can be solved by Pascal's triangle, and demonstrate its solution.

Sample Group Activities

Ch. 3: Connections: The Golden Ratio and Fibonacci Numbers

Names:

The Fibonacci sequence appears in patterns connected with the Golden Ratio ϕ. Conjecture: When ϕ is raised to a positive integer power, the result can be written as $A + B\phi$ where A and B are Fibonacci numbers

For example $\phi^2 = [1/2 + (1/2)\sqrt{5}][1/2 + (1/2)\sqrt{5}] = 3/2 + (1/2)\sqrt{5}$. I want to find two Fibonacci numbers, A and B so that

$$\phi^2 = A + B\phi$$
$$3/2 + (1/2)\sqrt{5} = A + B[1/2 + (1/2)\sqrt{5}]$$
$$= A + (1/2)B + (1/2)B\sqrt{5}.$$

If $A = 1$ and $B = 1$ (the first two Fibonacci numbers), we get

$$3/2 + (1/2)\sqrt{5} = 1 + 1[1/2 + (1/2)\sqrt{5}]$$
$$= 1 + (1/2)(1) + (1/2)(1)\sqrt{5}$$
$$= 3/2 + (1/2)\sqrt{5}$$

Find ϕ^3; ϕ^4 to gather more evidence for this conjecture.

Can you determine a pattern emerging from your calculations? That is, try to generalize this result for ϕ^n, using the Fibonacci sequence and writing A and B as u_n for some n. Hint: Begin by writing ϕ^2, ϕ^3, ϕ^4 in terms of the following:

$$1,\ 1,\ 2,\ 3,\ 5,\ 8,\ 13, 21, 34, 55, 89, \ldots, u_n, \ldots$$
$$u_1, u_2, u_3, u_4, u_5, u_6, u_7, u_8, u_9, u_{10}, u_{11}, \ldots, u_n, \ldots$$

Companion Guide to The Mathematical Experience

Ch. 7: From Certainty to Fallibility

Names:

Write a short paragraph in response to each of these questions. Assume that your audience has read *Platonism, Formalism, and Constructivism,* has attended class and seen "The Story of π," and "The Pythagorean Theorem."

1. Platonism views mathematics as a study about eternally given objects which therefore impose their nature upon the results of Mathematics. Mathematics thus deals with abstract or ideal "objects" which have *being* independent of our thought about them. How would the Platonist talk about π? That is, how would the Platonist explain why π can be found in number theory, probability, geometry, etc.?

2. Constructivists admit only the existence of mathematical objects and theories that can be constructed. Constructivists emphasize the *algorithmic* aspects of mathematics. Why do constructivists deny Cantor's continuum hypothesis?

3. Formalists say there are no mathematical objects—just formulas. What would the formalist say about the fact that the Pythagorean theorem has applications in the physical world? Does their response have some connection with the view of mathematics by fiat?

Sample Group Activities

Ch. 8: Mathematical Reality

Names:

Now that you have seen the videotape "The Hypercube," answer the following questions.

Different pictures of a 4-dimensional cube are mutually contradictory if we think of it as a 3-dimensional object. In 4-dimensional space, the different 3-dimensional projections fit together.

1. (a) Describe the *revolving door illusion* when you rotate a 3-cube in 3-space.

1. (b) Describe the *revolving door illusion* when you rotate a 4-cube in 4-space.

2. (a) How many corners does a square have?

2. (b) How many corners does a 3-cube have?

2. (c) How many corners does a 4-cube have?

3. (a) Describe what we mean by *perspective distortions* of a 3-cube.

3. (b) Describe what we mean by *perspective distortions* of a 4-cube.

4. (a) How many square faces does a 3-cube have?

4. (b) How many cubical faces does a 4-cube have?

4. (c) How many edges come out of each corner of a 3-cube

4. (d) How many edges come out of each corner of a 4-cube

5. (a) What geometric figures do you obtain when you *slice* a square, *corner first*, with a *one-dimensional knife?*

5. (b) What geometric figures do you obtain when you *slice* a 3-cube, *corner first*, with a *two-dimensional knife?*

5. (c) What geometric figures do you obtain when you *slice* a 4-cube, *corner first*, with a *three-dimensional knife?*

6. (a) How is a square related to a 3-cube?

6. (b) How is a 3-cube related to a 4-cube?

Part V
Sample Examinations

Sample Examinations

Sample Take-Home Examination

I. (35 points) Answer each part completely. Descriptions should be at least one paragraph. Where appropriate, cite specific mathematical examples to support your answer.

a. What is a frieze pattern? What role does mathematics play in categorizing frieze patterns? Be sure to define specific mathematical terms that apply to frieze patterns. Create a frieze pattern of your own and specify its type.

b. What is the difference between a conjecture and a proof? Give an example of a mathematical conjecture. Describe two different types of mathematical proofs.

c. Describe the real number system. Be sure to include at least four different subsystems of numbers in the system. Give an example of a subset with an operation that forms a group within the real number system.

d. Without using a calculator, determine how much is $1 + 3 + 5 + 7 + \cdots + 999$? Explain how you found the sum.

e. Describe the symmetries of an equilateral triangle.

II. (20 points) Answer the following questions for this operation table:

*	a	b	c	d
a	a	b	c	d
b	b	d	a	c
c	c	a	d	b
d	d	c	b	a

1. Is this set closed under * ? Why or why not?

2. Is there an identity? If so, what is it? If not, why not?

3. Does every element have an inverse? If so, list each element with its inverse. If not, show which elements have no inverses and explain why not.

4. Give an example to show this set has associativity under *.

5. Does this table describe a group? Why or why not?

III. (15 points) Let $R = \{q, r, s, t\}$ with an operation $\sharp$ defined on R. Furthermore, assume the following properties:

q is the identity element

r is the inverse of s

$r \sharp r = s \sharp s = t$

$R, \sharp$ is a group

Construct an operation table for $R, \sharp$.

IV. (25 points) We have discussed patterns, groups, and proofs in this class. Choose one of these topics and write a two-page (typed, double space) magazine article for a general audience. Describe what you have learned about your topic in this class. Give your reader a sense of what your topic involves, by means of definitions and examples. Discuss the beginnings of your topic or place it in a historical or conceptual context if you can. Illustrate its applications. Describe it within the framework of issues we have discussed or read about: the growth of knowledge in mathematics; mathematics as invention or discovery; the mathematical community; mathematics as art or science, etc. Create a banner headline for your article that enables your reader to determine your goals in exposition.

Sample Examination

I. (40 points) Write any ONE of the essays specified below. Consider your audience to be students who are planning to enroll in this class in the Fall semester. You may assume they have read *The Mathematical Experience*. However, do not assume any knowledge of the mathematics you describe. To that end, be sure you give thorough explanations of the mathematical facts used in your essays.

A. Write a one-page essay based on this reading from *The Mathematical Experience* in response to the questions posed:

Confessions of a Prep School Teacher

I propose that Williams makes statements that are misleading, implying some things about mathematics that are not so. Example: "He prefers to teach mathematics rather than physics because it is hard to keep up with the new developments in physics." This can be interpreted by the reader as: there are no (or few) new developments in mathematics. As many people believe mathematics is a "collection of facts," this kind of implication feeds the fallacy. In other places, Williams responds with platitudes about mathematics, and then when questioned further, explains with convoluted reasoning or inaccurate definitions.

Give an illustration of the above. Then explain what is wrong with Williams' explanation. Use at least one specific mathematics fact to illustrate your point.

B. Write a one-page essay based on this reading from *The Mathematical Experience* in response to the questions posed:

The Classic Classroom Crisis of Understanding and Pedagogy

Explain what the authors mean by "fiddling around" and describe under what circumstances and to what purpose should a professor "fiddle around" in class? Based on your experience in this class, recount a specific episode of how "fiddling around" by your or your instructor assisted you in understanding a concept. Be sure your explanation describes the mathematics involved in the event.

C. Write a one-page essay based on this reading from *The Mathematical Experience* in response to the questions posed:

Polya's Craft of Discovery

Have you used, observed your instructor or your classmates use, any of the heuristics that Polya cites? Choose one problem. Explain the mathematics involved, and what heuristics were used to solve it.

II. (15 points) Derive a number sentence from figurate numbers, and then use mathematical induction to prove it.

III. (15 points) Define and give an example of each of the following:

  a. A counterexample.

  b. A primary Pythagorean triple.

  c. A conjecture in mathematics.

IV. (30 points) Answer each of the following with a one-paragraph response. Give a specific example for each of the six given themes which characterize the evolution of mathematics:

  a. Results may be accepted before proof can be demonstrated.

  b. One proof may not be enough.

  c. Problems generate new problems

  d. A change in context alters an accepted result.

  e. New branches of mathematics can be the consequence of investigating familiar facts.

  f. A discovery in one field can precipitate major crises in the mathematical world and beyond.

Sample Examinations

Sample Midterm Examination

1. (20 points) Find reasons for thinking that any brief definition of mathematics must be inadequate. Discuss your findings with someone who thinks that mathematics is simply number-crunching. Give specific examples to support your statements.

2. (20 points) Mathematics builds on itself. Branches of mathematics borrow from each other. Often we find discoveries in one branch apply to another branch. Estimating, conjecturing, proving, finding patterns are all activities that engage mathematicians. Discuss either the Pythagorean theorem, the number π, or the Fibonacci sequence and try to show how its history conveys specific attributes of mathematics.

3. (20 points) Answer each of the following questions and where possible, give an example to support your answer. If you give the name of a theorem, you must state the theorem:

a. Every provable statement is true but not every true statement is provable. What famous mathematician discovered this result? Give an example that possibly illustrates it.

b. Describe the role that counterexample plays in proving or disproving a theorem. Give an example.

c. What is the difference between conjecture and proof? Give an example of each.

d. Mathematicians try to find order out of chaos. Give an example from mathematics of this activity.

e. When little Gauss was in kindergarten his teacher gave him a mathematics problem to keep him busy. What was the problem, and how did Gauss solve it in record time (i.e., demonstrate Gauss' solution)?

4. While driving home from work at rush hour you must pass through eight intersections controlled by stoplights. Assume when you reach an intersection that the light is either green or red. Use Pascal's triangle to solve this problem:

a. In how many ways can you go through the intersections catching at least four green lights?

b. In how many ways can you go through the intersections in which all the lights are green?

c. In how many ways can you go through the intersections in which no more than two of the lights are green?

5. (20 points) Make a conjecture about the sum of Fibonacci numbers with odd subscripts and prove your conjecture using mathematical induction.

Sample Final Examination

In-class exam with one take-home problem. Each problem is worth 20 points.

1(A). Find the sum of the first 500 odd numbers, using the following procedures:

a. Find the sums of the first few odd numbers, noting totals, and expressing these totals in terms of exponents. Show at least four examples.

b. Describe *in words* the pattern you find, i.e., use a sentence to express the pattern in terms of an arbitrary number n.

c. Use your pattern to predict the sum of the first 500 odd counting numbers. Express the sum in terms of exponents.

1(B). Using the problem-solving technique of Carl Friedrich Gauss, find the following sum. You may express your answer as a product of two integers.

$$4 + 8 + 12 + \cdots + 396 + 400$$

2. Let $u_1, u_2, \ldots, u_n, \ldots$ be the Fibonacci sequence. Prove, for all positive integers n, that

$$(u_{n+1})(u_{n+2}) - (u_n)(u_{n+3}) = (-1)^n$$

3. You are studying the voting record of your senator. She voted on eight pieces of legislation. Use Pascal's triangle to answer the following questions.

a. How many different ways could she have voted?

b. How many different ways could she have voted *no* on at least three of the bills?

c. How many different ways could she have voted *yes* on no more than six of the bills?

4(A). Use the *figurate numbers* of the Pythagoreans to make a conjecture about sums of integers.

a. Describe your conjecture *in words*.

b. *Draw the figures* to support your statement.

4(B). The Fibonacci sequence appears in patterns connected with the golden ratio ϕ, when ϕ is raised to a positive integer power. Find ϕ^2; ϕ^3; ϕ^4; etc. to determine the pattern, i.e., write ϕ^n in terms of ϕ and Fibonacci numbers.

5. Are there any numbers common to the following three arithmetic sequences? If so,

a. construct a system of simultaneous congruences;

b. indicate why this system has a solution;

c. if it has a solution, *use the Chinese remainder theorem* to find the first three numbers common to the sequences.

If not, indicate why not.

Sequence #1: 5, 8, 11, ...
Sequence #2: 9, 13, 17, ...
Sequence #3: 2, 7, 12, ...

6. Take-home problem (to be submitted with the final examination): Assume you have been the instructor in this class. The chair of the department asks you to submit a report on one of your themes and its objectives and how you attempted to meet them.

Write such a report. Include:

1. *The theme you would have addressed and a catalogue of each topic under that theme.*

2. *Give* several *examples of how you would have demonstrated the theme with specific discussion of classroom activities and readings*

from the text. Here, you should convince the chair, *by specific examples*, of *how* the topics helped you demonstrate the theme.

3. *Indicate why you do or do not believe you have met your objective for the theme.* Here, think about whether you "grew" in your understanding of the theme by considering specific topics. It's not necessary to be positive in response to this. But your response, positive or negative, needs to be supported with specifics.

4. *Report what (if anything) you would change should you teach this class again and why, and/or give advice to other instructors who plan to teach the class.* This report should be approximately three to four pages. It needs to include *specific mathematical problems and citations from the text,* with interpretation of "lessons learned" by examining the solutions to the problems and the readings from the text. Name dropping and vague statements are discouraged. If you discuss Goldbach's conjecture, you must state the conjecture and give examples to demonstrate it. If you make the connection between Pascal's triangle and the binomial theorem, you must demonstrate problems with solutions! If you believe a topic appropriately demonstrated a theme, explain why with examples. If you cite a problem from a group project, you must give its solution and explain why, in developing the solution, your understanding of the theme was increased.

Part VI

Topics for Expository Research Papers

Topics for Expository Research Papers

1. "The definition of mathematics changes" (page 8 of text). Demonstrate this phenomenon. For example, trace the changing definition of geometry.

2. "Euclidean geometry can be defined as the science of ruler-and-compass construction" (page 13 of text). Use this definition to explain the three famous unsolved problems of antiquity.

3. Demonstrate how science and technology have been "sources of new mathematical questions" (page 25 of text).

4. In *Studies in the History of Mathematics* (published by The Mathematical Association of America), Esther Phillips collects recent articles in the history of algebraic number theory, geometry, topology, logic, and the relationship between mathematics and computing. Choose one of these topics and write a paper describing the significance of the research that is being done. Be sure to have at least two other sources to support your description.

5. David Eugene Smith's *Source Book in Mathematics* (McGraw Hill, 1929) attempts to present "the most significant passages from the works of the most important contributors to the major sciences from the 16th to the 19th centuries." The book is divided into sections on number theory, algebra, geometry, probability, and calculus. Choose one of these topics to explore. Include not only what Smith reports, but add results from more current research in the field. Consult other source books in mathematics, e.g., *A Source Book in Classical Analysis*, Garrett Birkhoff, ed. (Cambridge: Harvard University Press, 1973), *The World of Mathematics*, James R. Newman, ed. (New York: Simon and Schuster, 1956), *A Collection of Modern Mathematical Classics*, Richard E. Bellman, ed. (New York: Dover, 1961), *The World Treasury of Physics, Astronomy, and Mathematics*, Timothy Ferris, ed. (Boston: Little Brown, 1991).

6. Write a paper on a mathematician who is currently active, reconstructing a list of her publications as well as describing the focus of her research. How does this mathematician compare to the Ideal Mathematician of Chapter 2?

7. Imagine you are asked to add a chapter to *The Mathematical Experience*, and do it!

8. Choose a problem that is currently unsolved. Trace its history and describe some of the results that have been obtained in the attempts to solve the problem. Discuss how activity surrounding this problem conforms or doesn't conform to the views of mathematics presented in *The Mathematical Experience*.

9. Trace the development of trigonometry from the Greeks to the nineteenth century. Choose one of the outer issues of Chapter 3 and relate it to your study of trigonometry.

10. Describe elementary plane geometry as Euclid would. Then demonstrate how this view of geometry has been influenced by (1) Platonism, (2) the Pythagorean Theorem, (3) the discovery of non-Euclidean geometry, (4) the nineteenth/twentieth century research in axiomatics. Contrast your ideas with the ideas expressed in Chapter 7.

11. Consider the following characterizations of mathematics given by Maxime Bôcher in "The Fundamental Conceptions and Methods of Mathematics" (*The American Mathematical Monthly*, December 1904, pp. 115–135):

> Mathematics is the science of quantity... of space and number. (p. 115)
>
> Mathematics [is] concerned with those conceptions which are obtained by direct intuition of time and space without the aid of empirical intuition. (p. 116)

Select either or both of these characterizations and write a paper that supports or challenges them. Include specific mathematical references that you explore (not just cite!).

12. In *Night Thoughts of a Classical Physicist* (Cambridge: Harvard University Press, 1982), compare Russell McCormmach's physicist with the physicist you meet in *The Mathematical Experience*. Describe the role of mathematics in these physicists' worlds.

13. In *Mathematics of Great Amateurs* (New York: Oxford University Press, 1990), Julian Coolidge describes the accomplishments of men and women who did not make mathematics their main work yet whose contributions to mathematics are of permanent value. Choose one such figure, describe her or his contributions to mathematics, and investigate how this person's work influences some area in mathematics today. Or address this claim: The work of amateur mathematicians can no longer be significant because of the complexity of the subject. Be sure to consult at least two other sources in your research.

14. There have been a number of historical periods in which the pursuit and production of mathematics has been at a very low ebb. Explore this phenomenon. Discuss what reasons might be found for the decline of interest in the subject.

15. Find out something about "virtual reality" and speculate on the extent to which its potentialities might lead to a new type of futuristic mathematics.

16. Historians have suggested that schools of art or literature such as mannerism, romanticism, and realism came to an end because they exhausted their internal possibilities. Discuss this with respect to particular subjects within mathematics.

17. Interview a lawyer, a doctor, a store manager, a carpenter, etc. Find out how and the extent to which they use mathematics. Ask them also whether they have in their own minds the image of an "ideal" lawyer, etc. Then interview a mathematician. Ask similar questions.

18. Write a paper on a famous axiom. Select some of the following topics to explore:

    a. Give a formal statement of the axiom. Determine if there are different statements (i.e., versions) of the axiom. For example, the axiom of Archimedes can be stated in terms of numbers and in terms of line segments.

    b. Trace the history of the axiom (briefly).

    c. Discuss any controversies about the axiom. For example, the axiom of choice, which is needed to establish parts of analysis, topology, and abstract algebra, was considered objectionable by a number of mathematicians, among them Hadamard, Lebesgue, Borel, and Baire. Hilbert's axiom of completeness was the subject of controversy and was replaced by Bernays with a linear completeness axiom.

    d. Determine if the axiom is used to establish any important results or prove any famous theorems. For example, Zermelo used the axiom of choice to prove the well-ordering theorem. Archimedes' postulate and the completeness axiom endow the line and circle, for example, with the kind of structure that enables us to give a rigorous proof for the construction of an equilateral triangle.

    e. Explore the possibilities that exist in denying the axiom. Can new systems of mathematics be created by assuming the negation of the axiom? For example, Max Dehn invented (through a deliberate application of the axiomatic method) a geometry in which the postulate of Archimedes is denied. Dedekind's axiom of continuity (in the environment of Euclidean geometry) loosely spoken insures that a line has no "holes" in it. If we deny this axiom, we can get a model called the *surd* plane, a plane that is used to prove the impossibility of trisecting every angle with a straightedge and compass. Bachmann has developed geometries without using axioms of betweenness or continuity.

    f. If applicable, discuss the view of the axiom by Platonists, formalists, or constructivists, and what impact, if any, the axiom had on activities of these schools. For example, the axiom of reducibility drew such severe criticism that the formalist school concentrated much activity in attempting to devise some method of avoiding it.

    g. Discuss motivations for introducing the axiom. For example, to guarantee the existence of certain points of intersection (of line with circle and circle with circle), Richard Dedekind introduced into geometry his continuity postulate: "If all points of a horizontal straight line fall into two classes, such that every point of the first class lies to the left of every point of the second class, then there exists one and only one point that produces this division of all points into two classes—that is, this severing of the straight line into two portions."

h. Is your axiom sometimes (often?, seldom?) taken as a theorem? For example, Giuseppe Peano postulates mathematical induction but Mario Pieri deduces it as a theorem.

i. Have axioms been proposed that contradict the axiom? For example, an axiom of determinacy (for certain games) in its full form contradicts the axiom of choice.

j. Does the axiom play a role in distinguishing between different branches of mathematics? For example, the parallel postulate figures significantly in distinguishing Euclidean geometry from non-Euclidean geometry.

k. You might show how the axiom is used to prove a certain theorem. For example, show how, as a consequence of Pasch's postulate, that one can prove if a line enters a triangle at a vertex, it must cut the opposite side. You might extend the axiom. For example: extend Dedekind's axiom to cover angles; restate Archimedes' axiom for angles and indicate how it might be deduced from the arithmetized form of the postulate. You might demonstrate where the axiom does not hold. For example, Pasch's postulate does not always hold for a spherical triangle cut by a great circle.

l. If applicable, determine if mathematical theories can be built up exclusively in terms of specific axioms. For example, Huntington devised a system based on axioms of betweenness.

m. If applicable, discuss any open questions that the axiom resolved, or engendered. For example, Zermelo's axiom of infinity guaranteed the existence certain infinite sets, but there was nothing in his system to guarantee the existence of the union of those sets. This was something Fraenkel later worked on, and as a result, proposed a new axiom: the axiom of replacement. The original axioms of Zermelo, amended by Fraenkel, came to be known as a new theory of sets: the Zermelo-Fraenkel axioms. Then, enter Von Neumann and .... There are many interesting mathematical questions that cannot be settled on the basis of the Zermelo-Fraenkel axioms for set theory.

n. Discuss equivalences to the axiom. For example, Pasch's postulate is equivalent to the separation axiom: A line m separates the points of the plane which are not on m into two sets such that if two points X and Y are in the same set, the segment XY does not intersect m, and if X and Y are in different sets, the segment XY does not intersect m. The axiom of choice is equivalent to Zorn's lemma. The Archimedean axiom is equivalent to the statement: The limit as **n** goes to infinity of $l/\mathbf{n}$ is equal to zero.

o. What does the axiom "buy" us? For example, axioms of betweenness assure us of the existence of an infinite number of points on a line and that a line is not terminated in any point. They guarantee us that the order of the points on a line is serial rather than cyclical. Pasch's postulate gives us information about the plane as a whole.

p. How does a change of environment affect the axiom? For example, are the continuity axioms for Euclidean geometry the same as the continuity axioms for projective geometry? Separation axioms? Order axioms?

q. What is the relationship between an axiom and reality?

19. The ancient Chinese had an extensive mathematics in which there were no proofs. Read Joseph Needham's *Science and Civilization in China* (Cambridge: Cambridge University Press, 1954) on this point, and discuss.

20. Is mathematics discovered or invented? Research this question with regard to different schools of thought in historical and contemporary mathematics. Take a position, and give specific examples to support your position.

21. Investigate calculus. Follow its genesis from antiquity. Describe its importance to mathematics. What problems was it developed to solve? What complex array of science now depends on it?

22. Mathematics has been called "the conquest of intuition." What is the nature and role of mathematical intuition? What controversies surround it?

23. *Mathemata* means "to learn." Show how general methods of learning are reflected in mathematics.

24. Investigate a case where some mathematics pursued in the abstract as an end in itself, later turned out to have practical applications in science.

25. Do a brief history of mathematical pedagogy. What ideas have influenced its evolution? How does it relate to mathemat-

Topics for Expository Research Papers

ical research? Is there a difference between mathematics as it is done and how it is taught?

26. Does mathematics belong in the sciences, in the humanities, or in both? Discuss and support your answer with specific examples.

27. Investigate the role of induction (*not* Mathematical Induction) in mathematical discovery. Contrast this with deductive methods of proof.

29. How has the computer influenced the face of mathematics? What kinds of proof can a computer achieve? In what directions will it lead mathematics?

30. Where do mathematical ideas originate? What is their source and what makes them grow?

Suggestions for Grading Essays and Research Papers

As mathematicians, we approached the task of grading essays with some reluctance. The first few semesters we struggled with the problems involved with this kind of subjective grading.

However, we discovered a method that has made me much more comfortable with grading essays and research papers. It is based on the philosophy that writing (like mathematics) is a process. This is the method:

1. Assign the essay and tell students they will have one opportunity for a rewrite.

2. Collect the essays, and read each one without any intention of assigning a grade. Simply read the essay and make comments, indicating weaknesses. For example, note logical lapses in the development, indicate where specific examples are needed to prove their points, etc.

3. Return the essays to the students.

4. When the students submit the rewrites, grade them primarily on how well they overcame the weaknesses in their papers.

This sounds like it may be more time consuming because you must read the essays twice, but we have found this not to